Haiti M_W 7.0 Earthquake of January 12, 2010

Lifeline Performance

EDITED BY
Curtis L. Edwards, P.E.

SPONSORED BY
Technical Council on Lifeline Earthquake Engineering
Monograph No. 35

ASCE

Published by the American Society of Civil Engineers

Cataloging-in-Publication data on file with the Library of Congress.

Published by American Society of Civil Engineers
1801 Alexander Bell Drive
Reston, Virginia 20191
www.asce.org/pubs

ISBN 978-0-7844-1240-4 (paper)
ISBN 978-0-7844-7695-6 (PDF e-book)
Manufactured in the United States of America.

18 17 16 15 14 13 12 11 1 2 3 4 5

TCLEE Monograph Series

These publications may be purchased from ASCE via telephone at 1-800-548-ASCE (2723) or online at www.asce.org/bookstore. The TCLEE Web site is www.asce.org/tactclee.

No. 1 Recent Lifeline Seismic Risk Studies, Kiremidjian, Anne S., Editor, 1990.

No. 2 Seismic Lost Estimates for a Hypothetical Water System, A Demonstration Project, Taylor, Craig E., Editor, August 1991.

No. 3 Guide to Post-Earthquake Investigations of Lifelines, Schiff, Anshel J., Editor, August 1991.

No. 4 Lifeline Earthquake Engineering, Proceedings of the 3rd U.S. Conference on Lifeline Earthquake Engineering, August 22-23, 1991, Los Angeles, CA, Cassaro, Michael, Editor, August 1991.

No. 5 Lifeline Earthquake Engineering in the Central and Eastern United States, Ballantyne, Donald, Editor, September 1992.

No. 6 Lifeline Earthquake Engineering, Proceeding of the 4th U.S. Conference on Lifeline Earthquake Engineering, August 10-12, 1995, San Francisco, CA, O'Rourke, Michael J., Editor, August 1995.

No. 7 Critical Issues and State of the Art on Lifeline Earthquake Engineering, Schiff, Anshel J. and Buckle, Ian, Editors, October 1995.

No. 8 Northridge Earthquake: Lifeline Performance and Post-Earthquake Response, Schiff, Anshel J., Editor, August 1995.

No. 9 Seismic Design for Natural Gas Distributors, McDonough, Peter W., August 1995.

No. 10 Methods of Achieving Improved Seismic Performance of Communications Systems, Tang, Alex, and Schiff, Anshel J., Editors, September 1996.

No. 11 Guide to Post-Earthquake Investigation of Lifelines, Schiff, Anshel J., Editor, July 1997.

No. 12 Seismic Guidelines for Ports, Werner, Stuart D., Editor, March 1998.

No. 13 Overcoming Barriers: Lifeline Seismic Improvement Programs, Taylor, Craig E., Mittler, Elliott, and Lund, Le Val, September 1998.

No. 14 Hyogo-Ken Nambu Earthquake of January 17, 1995—Lifeline Performance, Schiff, Anshel J. Editor, 1998.

No. 15 Guidelines for the Seismic Evaluation and Upgrade of Water Transmission Facilities, Eidinger, John M. and Avila, Ernesto A., Editors, January 1999.

No. 16 Optimizing Post-Earthquake Lifeline System Reliability (Proceedings of the 5th U.S. Conference on Lifeline Earthquake Engineering, Seattle, Washington, August 12-14, 1999.), Elliott, William M., and McDonough, Peter W., Editors, August 1999.

No. 17 Ismit (Kocaeli), Turkey Earthquake of August 16, 1999, Including Duzce Earthquake of November 12, 1999—Lifeline Performance, Tang, Alex K., Editor, September 2000.

No. 18 Chi-Chi, Taiwan, Earthquake of September 21, 1999—Lifeline Performance, Schiff, Anshel J., and Tang, Alex K., Editors, October 2000.

No. 19 Gujarat (Kutch) India, M7.7 Earthquake of January 26, 2001 and NAPA M5.2 Earthquake of September 3, 2000, Eidinger, John M., Editor, June 2001.

No. 20 The Nisqually, Washington, Earthquake of February 2001—Lifeline Performance, McDonough, Peter W., Editor, February 2002.

No. 21 Acceptable Risk Process—Lifelines and Natural Hazards, Taylor, Craig E., and VanMarcke, Erik H., Editors, March 2002.

No. 22 Seismic Screening Checklists for Water and Wastewater Facilities, Heubach, William F., Editor, September 2002.

No. 23 Atico, Peru Mw 8.4 Earthquake of June 23, 2001, Curtis L. Edwards, Editor, October 2002.

No. 24 Lifeline Performance of El Salvador Earthquakes of January 13 and February 13, 2001, Le Val Lund, Editor and Carl Sepponen, Editor, September 2002.

No. 25 Advancing Mitigation Technologies and Disaster Response for Lifeline System: Proceedings of the Sixth U.S. Conference and Workshop on Lifeline Earthquake Engineering, Beavers, J.E., Editor, August 2003.

No. 26 Fire Following Earthquake, Scawthorn, Charles, Eidinger, John M., and Schiff, Anshel J., Editors, 2005.

No. 27 Zemmouri, Algeria, Mw 6.8 Earthquake of May 31, 2003, Edwards, Curtis L., Editor, 2004.

No. 28 San Simeon Earthquake of December 22, 2003 and Denali, Alaska, Earthquake of November 3, 2002, Yashinsky, Mark, Editor, Lund, Le Val, Co Editor, 2004.

No. 29 Hurricane Katrina: Performance of Transportation Systems, DesRoches, Reginald, Editor, 2006.

No. 30 Sumatra-Andaman Islands Earthquake and Tsunami of December 26, 2004 Lifeline Performance, Strand, Carl and Masek, John, Editors, 2007.

No. 31 Kashiwazaki, Japan Earthquake of July 16, 2007 Lifeline Performance, Tang, Alex K.K. and Schiff, Anshel J., Editors, 2007.

No. 32 Pisco, Peru Earthquake of August 15, 2007 Lifeline Performance, Tang, Alex K.K. and Johansson, Jorgen, Editors, 2007.

No. 33 Pacific Northwest Storms of December 1-4, 2007: Lifeline Performance, edited by Teresa Elliott, P.E., and Alex K. Tang, P.E.

No. 34 Seismic Resilience of Natural Gas Systems: Improving Performance, edited by Peter W. McDonough, P.E.

No. 35 Haiti M_w 7.0 Earthquake of January 12, 2010: Lifeline Performance, edited by Curtis L. Edwards, P.E.

Other TCLEE Publications

Duke, C. Martin, Editor, The Current State of Knowledge of Lifeline Earthquake Engineering, Proceedings TCLEE Specialty Conference, August 30-31, 1977, Los Angeles, CA, (Later designated as the 1st U.S. Conference on Lifeline Earthquake Engineering), August, 1977.

Dowd, Munson, Editor, Annotated Bibliography on Lifeline Earthquake Engineering, 1980.

Smith, D. J, Jr., Editor, Lifeline Earthquake Engineering, The Current State of Knowledge 1981, Proceedings of the Second TCLEE Specialty Conference, August 20-21, 1981, Oakland, CA, (Later designated as the 2nd U.S. Conference on Lifeline Earthquake Engineering), August 1981.

Hall, William J., Advisory Notes on Lifeline Earthquake Engineering, 1983.

Nyman, Douglas, NSF Principal Investigator, Guidelines for the Seismic Design of Oil and Gas Pipelines Systems, TCLEE Committee on Gas and Liquid Fuels, 1984.

Cooper, James, Editor, Lifeline Earthquake Engineering Performance, Design and Construction, 1984.

Cassaro, Michael and Martinez-Romero, E., Editors, The Mexico Earthquake, 1985, Factors Involved and Lessons Learned, 1986.

Eguchi, Ronald and Crouse, C. B., Lifeline Seismic Risk Analysis - Case Studies, 1986.

Wang, Leon R. L. and Whitman, Robert, Seismic Evaluation of Lifeline Systems-Case Studies, 1986.

Cassaro, Michael and Cooper, James, Editors, Seismic Design and Construction of Complex Civil Engineering Systems, 1988.

Werner, Stuart D. and Dickenson, Stephen E., Editors, Hyogo-Ken Nambu (Kobe) Earthquake of January 17, 1995: A Post-Earthquake Reconnaissance of Port Facilities, TCLEE Ports Committee, 1996.

ASCE Manual

Schiff, Anshel J., Editor, Guide to Improved Earthquake Performance of Electric Power Systems, ASCE Manual 96.

TCLEE Earthquake Investigation Reports

TCLEE has also prepared numerous earthquake reports that have appeared in other publications. References to these reports and 10 short reports associated with TCLEE monographs can be viewed on the ASCE/TCLEE Web site address given below. The 10 short reports are each about 5 to 15 pages long, contain a summary of main observations and some pictures. They can be downloaded at *www.asce.org/tactclee.*

Contents

Preface

The Earthquake Investigations Committee of the Technical Council of Lifeline Earthquake Engineering (TCLEE), American Society of Engineers (ASCE), was established to initiate, organize, train for, coordinate, and evaluate the performance of lifelines following earthquakes. Members of the committee are employees of lifeline industries, consulting engineers, and academics from the United States, Canada, Europe, and Asia. Committee members provide services on a voluntary basis. Travel expenses are covered in part by ASCE. Some members' companies may allow for special paid leave and may also help cover travel costs. In addition to the time and expense of the field investigation, members must spend several weeks of their time preparing the reconnaissance report.

Individuals participating in the investigation need not be members of the committee or ASCE but are expected to follow the committee's earthquake investigation practices as described in the ASCE publication *Guide to Post-Earthquake Investigation of Lifelines* and the July 25, 2009 *ASCE Post-Disaster Assessment Manual.* Members of the investigation team coordinate with other groups and may participate in groups organized by other organizations. They gather performance data, both good and bad, from earthquakes to provide information to help practitioners improve the performance of lifeline systems. The international earthquakes that have been investigated by this committee include 1985 Chile, 1988 Soviet Armenia, 1990 Philippines, 1991 Costa Rica, 1992 Turkey, 1995 Kobe, 1999 Turkey (Izmit), 1999 Chi Chi (Taiwan), 2001 El Salvador, 2001 Peru, 2004 Algeria, 2004 Sumatra, 2007 Kashiwazki Japan, and 2010 Chile. Domestic earthquakes investigated include 1989 Loma Prieta, 1992 Landers, 1994 Northridge, 2000 Napa, 2001 Nisqually, 2002 Denali, and 2003 San Simeon. TCLEE also conducted an investigation of 2006 Hurricane Katrina.

The Earthquake Investigations Committee works closely with the Earthquake Engineering Research Institute (EERI) to provide an abbreviated version of lifeline performance in *Earthquake Spectra*, an EERI publication.

Curtis Edwards, Editor
June 2011

Contributors

Curtis Edwards, *Editor*: Introduction, Chapters 1, 6, and 9
Pierre Alex Augustin: Chapter 7
Don Ballantyne: Chapter 3
Bill Bruin: Chapter 7
Rick Carter: Chapters 4 and 5
Brucely Joseph: Chapter 3
Aimee Lavarnway: Chapters 2 and 9
Nason McCullough: Chapter 7
Mark Pickett: Chapter 8
Dave Plum: Chapter 3
Stu Werner: Chapter 7

Due to the nature of post-earthquake investigations, this report may contain inaccuracies or be incomplete, notwithstanding efforts to eliminate errors.

1. Introduction

1.1 Earthquake Highlights

General

On January 12, 2010, at 4:53 p.m. local time, there was a magnitude 7.04 Mw earthquake in the western peninsula of the Republic of Haiti. It is estimated there were more than 250,000 deaths, 300,000 injuries, and 1,200,000 people left homeless and in temporary shelters. There were about 105,000 homes, 1300 schools, and 50 hospitals destroyed. More than 5 million people live in the earthquake area, which greatly challenged the government and relief agency efforts to restore basic housing and services. Additionally, many infilled walls failed on reinforced concrete buildings, and some engineered reinforced concrete structures failed.

Seismic Setting and Geotechnical Effects

Haiti is located on the western one-third of the island of Hispaniola in the Caribbean Sea. The island is situated in the north central region of the Caribbean plate near the boundary with the North American plate. The current understanding of the region's tectonic activity has generally been developed within the past 30 years and was still evolving at the time of the January 12, 2010 moment magnitude (M_w) 7.0 earthquake. The relative movement between the Caribbean and North American plates is somewhat more complex than some other plate boundaries. The January 12, 2010 earthquake occurred at 4:43:10 p.m. local time at 18.457° N 72.533° W at a depth of 13 km and had a moment magnitude (M_w) of 7.0 (USGS 2010). It was characterized by primarily left-lateral strike-slip movement propagating from east to west (Hayes 2010).

There was a great deal of damage to the port and coastal roads due to liquefaction and lateral spreading. In addition numerous landslides damaged homes and blocked roads.

Water and Wastewater Systems

Numerous water systems serve Haiti, including both public and private systems. The main public system in Port-au-Prince is supplied by a series of springs located in the nearby mountains. The water is chlorinated in the spring boxes and sent to the distribution system, which serves 250,000 connections and 1 million people. This water was not drinkable without further treatment before the earthquake and was not a reliable supply. Users typically have cisterns on their properties and augment the supply by purchasing water from private water trucks or by using collected rainfall. Homes not connected to the public system obtain their water from water trucks, which distribute treated well water to customers. There also is extensive use of bottled water. There were relatively few water main breaks, which is unusual for this large a system. Most breaks were repaired within 1 or 2 weeks. Unfortunately, the earthquake caused the loss of 50 percent of the water systems' paying customers.

After the earthquake, the U.S. military and non-governmental organizations (NGOs) set up water treatment plants to augment the private water supplies. This water was also distributed by trucks. Some of these systems were located near tent cities that had been set up as refuges for displaced Haitians.

There are no working wastewater treatment plants in Haiti. In the metropolitan areas, wastewater is discharged in open drainage channels and directed to the bay. In other areas, homes and businesses use septic tanks and pits to dispose of sewage. Many of these were damaged due to land movement causing pipe separation. Additionally, many of the urban drainage channels were blocked by building debris and trash. At the time of our visit, the government was working to clear the channels before the rainy season. The camps set up for displaced Haitians had rudimentary sewage systems, and some had chemical toilets. Waste from these locations was collected and taken to a landfill or spread on an open field, causing a significant health threat for the hundreds of thousands of people being exposed to contaminated water.

Highway Systems

The Haitian road and highway systems were not the best before the earthquake. After the earthquake, urban roads were blocked with debris, and limited equipment allowed for clearing only enough for one-way traffic on some streets. In coastal areas, many highways were damaged due to lateral spreading from liquefaction. These damaged roads had not been repaired and were barely passable. There were no reports of bridge failures, although some damage was noted. Other roads were damaged or blocked by numerous landslides in the region. Some were being cleared during our inspection, but many were not and were reduced to one-way traffic.

Electric Sector

We called on the National Rural Electric Cooperative Association's (NRECA) Myk (Mike) Manon, the association's international branch representative in Haiti, to assist Electricité d'Haïti (EDH) with system restoration. As we travelled to one of the major Port-au-Prince (PAP) substations, Manon provided an overview of the Dominican Republic power authority's effort to restore service to this substation, where equipment was literally knocked out of service. Notably, we discussed the progress and the challenges to system recovery and its return to normal business operations. Manon mentioned that the EDH system layout is not recorded on maps, and as a result, the initial response and recovery efforts were reliant on the memory of the line management and workforce. Additionally, the EDH system is not supported by SCADA or any other comparable systems management protocol.

The TCLEE team visited a diesel power generation facility close to the epicenter, which experienced severe building damage and geologic subsidence. The recovery of the facility and its generation capability has yet to be determined. The Peligre dam-hydro facility was constructed in the early 1950s, primarily for flood control and water supply and then modified in the 1970s for power. This facility was not damaged and its ability to generate power was not impaired, though power generation was temporarily halted to isolate damaged areas of the feeder and distribution systems downstream. Transmission lines were said to have been stable and functional through the incident.

Telecommunications Sector

We attempted contact with the government telecommunications public switched telephone network operator, Teleco, but did not receive a reply to our inquiries. During the earthquake the building that housed the access tandem switch and the trans-Caribbean Sea fiber optic landing site was destroyed.

By the time of our field visit, the rubble from the multi-story collapse had been cleared, NS we had an opportunity to view the cable vault, which was extending into the public right of way, and the sub-basement cable raceways, which seem to be intact. A further investigation by the operator is necessary to review the functionality and feasibility of facility restoration.

Digicel, one of the Haitian wireless operators, experienced a short-term service interruption when its HVAC and environmental system protocols were damaged, precipitating the shutdown of its equipment. A rapid work around of the building HVAC redirected to the equipment room allowed service restoration, even though the office building sustained some structural damage. According to a briefing provided through the National Communications System, a division of the Federal Department of Homeland Security (www.ncs.gov), the wireless systems were back to their pre-incident capability about 2 weeks after the January 12 event.

In summary, the overhead power and telecommunications systems, including CATV systems, collectively sustained significant damage as a result of an adjacent building collapse. Fortunately, the systems were not impaired in all the areas affected by the earthquake. The one exception is the collapse of the Teleco building and the destruction of the access tandem switch, which only halted inter-office communication and international wire-line telecommunications. The end offices were still able to send voice and data downstream through the outside plant facilities to the intended end-user. The re-establishment of interoffice traffic was accomplished by various means, which are discussed in the Chapter 5.

Substantial areas of the outside plant experienced years of deferred maintenance and in some cases poor construction methods and/or a lack of skilled and certified craftsmen complying with an acceptable national or international construction standard.

Ports

Autorite Portuaire Nationale (APN) operates several facilities in Port-au-Prince. It is the largest and busiest container port in Haiti, handling about 1200 containers per day and 170,000, 20-ft. equivalent units (TEUs) per year before the earthquake, according to APN officials. The port consisted of two separate waterfront facilities, designated as the North Wharf and the South Pier. The North Wharf structure is a pile-supported marginal wharf 1500 ft. long and was completely underwater after the earthquake. A gantry container crane and a mobile container crane along the North Wharf were partially submerged due to the collapse of the wharf structure. The South Pier is a 1250-ft.-long finger pier that is connected to a small island by two small bridges extending perpendicular to the pier. The pier was heavily damaged during the earthquake, with the western-most 400 ft. of the pier collapsing and 90 percent of the remaining piles requiring repair of the pile-deck connection. The port's main access road was heavily damaged by lateral spreading due to liquefaction of the foundation soils.

Recovery efforts to restore port operations began 2 days after the earthquake and were carried out as an around-the-clock U.S military task force operation. The U.S. military initiated efforts to repair the uncollapsed segment of the South Pier, including the construction of a reinforced concrete cap, which encased the top few feet of the severely damaged piles.

Varreux Terminal is a private multi-use terminal located north of Port-au-Prince. It consisted of a marine oil terminal, two piers, petroleum storage tanks, and a truck loading facility that supported the marine oil terminal. The terminal's western-most berths consisted of a concrete pier about 240 ft. long by 25 ft. wide supported on a steel substructure with concrete-filled steel pipe piles. The above-water steel substructure and pile were observed during the investigation to be severely corroded. The observed pier damage was attributed to inertial pounding between two portions of the deck during the earthquake, but the pier remained in operation following the earthquake. The eastern-most berths consisted of a steel pile-supported concrete pier about 560 ft. long and 80 wide on the north end, connected to a 30-ft.-wide marginal wharf backfilled with soil on the south end. The pier (≈345 ft. long) collapsed during the earthquake causing 56 casualties, and the remaining marginal wharf suffered several feet of lateral movement attributed to liquefaction-induce lateral spreading. There was also damage to a storage warehouse adjacent to the terminal. Only limited information was available, but it appeared that portions of the warehouse had collapsed during the earthquake. Several tanks in the terminal tank farm experienced mid-height buckling as well as elephant-foot buckling at their base. There was no indication of failure of the damaged tanks. Following the earthquake, the terminal's operations were restored with only minor downtime. Petroleum was being offloaded during our March 1, 2010 visit.

The Thor Marine Oil Terminal is a private terminal located southwest of Port-au-Prince. The terminal included an earth embankment extending about 800 ft. out from the shoreline, beyond which a 400-ft.-long concrete pile-supported pier extended. The earth embankment underwent significant liquefaction and lateral spreading with vertical displacements of up to 5 to 6 ft. and lateral movements of 10 ft. or more. The pipelines supported on the embankment were damaged and became unusable after the earthquake. There was noted inertial damage to the pile-supported pier, with some deck slabs having up to 2 ft. of permanent offset from the pile caps. Two tanks were damaged during the earthquake, undergoing mid-height and elephant-foot buckling. One of the tanks failed and lost all its diesel contents, which appeared to have failed at a lap weld of the bottom plate. The offloading pipeline was not in operation after the earthquake. However, the next ship to call brought in floating lines, which were used to offload petroleum products. These floating lines have subsequently been used to offload ships and will replace the embankment and pier for the near future. There are plans to repair the damaged tanks; plans to repair the embankment and pier are unknown.

Hospitals

To coordinate international response to the earthquake, the Haitian Ministry of Health has defined three levels of healthcare facilities—mobile health centers, fixed health centers for patients with minor problems, and hospitals with surgical capacities. The World Health Organization (WHO)/Pan American Health Organization (PAHO) has compiled a list of more than 900 healthcare facilities in Haiti. In the aftermath of the earthquake, there are about 91 functioning hospitals, including 59 in the metropolitan Port-au-Prince area. In this metro area, functioning facilities include four public, 34 private (providing free care since the earthquake), and 21 field hospitals. Fifty-six of the 59 hospitals have surgical capacity. The Health Ministry lost more than 200 staff members due to the earthquake.

All of the facilities investigated reported that their emergency generators were operated on a daily basis before the earthquake because of the intermittent nature of the commercial electrical power supply. Similarly, it was reported that hospitals always purified the piped-in commercial water due to the intermittent nature of the commercial water supply. There were no reports of problems with the wastewater system.

It was reported that patients were reluctant to enter hospital buildings due to fear of collapse during aftershocks. Consequently, almost all healthcare services were provided in tents even 7 weeks after the event. All facilities (public and private) were providing services to patients free-of-charge. At least one private hospital could no longer pay its staff because of the lack of funding. Both public and private hospitals were eager for written recommendations on seismic retrofit and upgrade. They intended to use the recommendations, along with data on patients served for free, in proposals to NGOs for future funding.

Other Lifelines

There is no natural gas system in Haiti. Gasoline and liquid fuels are transported by truck. Deliveries suffered minor delays due to highway damage.

1.2 Lessons Learned

This monograph highlights the damage and impacts to lifelines in Haiti. The lessons learned include some atypical impacts on lifelines from this earthquake.

1. As with other earthquakes, unreinforced masonry structures continue to cause lifeline damages during earthquakes. Simple reinforcement measures and use of reinforced concrete framing can greatly improve earthquake performance.
2. The stability of excavated slopes should be evaluated. Where steep, unstable slopes loom above lifelines, they should include walls and/or fencing to contain and/or deflect debris away from lifeline facilities.
3. The lack of available supplies and equipment is causing major delays in the full restoration of lifelines in Haiti.

4. The lack of reliable clean water supplies and the absence of wastewater collection and treatment systems is greatly increasing the risk of disease and other health issues.
5. Roads were mostly damaged by landslides and liquefaction. Many roads were damaged due to subsidence of poorly compacted fills and blockage by landslides and rock falls.
6. Most people in Haiti have a cell phone instead of a landline. Although there was damage to cell towers, the cellular providers restored service using temporary towers and temporary power.

1.3 Investigation Team

The information presented in this report was developed based on a field reconnaissance in Haiti conducted by a team fielded by the Technical Council on Lifeline Earthquake Engineering (TCLEE), which is a committee of the American Society of Civil Engineers (ASCE). The team consisted of the following:

Mr. Curt Edwards (Team Leader, Psomas, San Diego, California)
Mr. Pierre Alex Augustin (Caltrans, Sacramento, California)
Mr. Stu Werner, (Seismic Systems, Oakland, California)
Ms. Aimee Lavarnway (Shannon Wilson, Portland, Oregon)
Mr. Dave Plum (Consultant, Virginia Beach, Virginia)
Mr. Rick Carter (State of Oregon, Portland, Oregon)
Mr. Nason McCullough (CH2M Hill, Corvallis, Oregon)
Mr. Don Ballantyne (MMI, Seattle, Washington)
Mr. Mark Pickett (Univ. of Toledo, Ohio)
Mr. Brucely Joseph (URS)
Mr. Bill Bruin (Halcrow, San Francisco, California)

The TCLEE team shared planning and resources with the investigation team sponsored by the Earthquake Engineering Research Institute (EERI). Without the hard work of EERI team leader Dr. Reggie DesRoches our investigation would not have been successful.

The people of Haiti provided great assistance to the TCLEE team, and without their cooperation and sharing of damage information, the findings presented in this report would not have been possible. We express a great gratitude to everyone who assisted us in our efforts.

All photos are courtesy of the ASCE TCLEE Investigation Team unless otherwise noted.

1.4 Web Addresses

This investigation would not have been a successful without the acquisition of useful information from the following Web sites:

www.asce.org
www.eeri.org
www.worldcell.com
www.abcnews.go.com
www.msnbc.com
travel.state.gov/haiti
www.cdc.gov/travel
www.cnn.com
www.usgs.gov
earthquake.usgs.gov/activity/latest/
www.alertnet.org
news.bbc.co.uk/
www.reliefweb.int/w/rwb/nsf
dailynews.yahoo.com
www.nytimes.com
www.usatrade.gov
www.tradeport.org
www.countrywatch.com
www.reliefweb.int

These sites should be contacted for information relating to travel, damage, and conditions for future Haitian earthquakes.

1.5 Abbreviations

ASCE	American Society of Civil Engineers
CMU	concrete masonry unit
g	acceleration of gravity (=32.2 ft./sec/sec or 981 cm/sec/sec)
gal	acceleration (986 gal = 1 g)
M_w	Moment magnitude
M	Magnitude (of aftershocks)
MMI	Modified Mercalli Intensity
MGD	Million gallons per day (U.S. measure)
PGA	Peak ground acceleration (g)
TCLEE	Technical Council on Lifeline Earthquake Engineering
V	Base shear, for design

2. Seismology and Geotechnical Issues

2.1 Seismology

Geology and Seismotectonics

Haiti is located on the western one-third of the island of Hispaniola in the Caribbean Sea. The island is situated in the north central region of the Caribbean plate near the boundary with the North American plate. The current understanding of the region's tectonic activity has generally been developed within the past 30 years and was still evolving at the time of the January 12, 2010 moment magnitude (M_w) 7.0 earthquake. The fault-bounded area between the Enriquillo-Plantain Garden fault zone (EPGFZ) on the south and the Septentrional fault zone (SFZ) to the north was suggested by Mann et al. (1995) as a developing microplate. Based on the general movement of the microplate, a portion of the Caribbean plate is in the process of being transferred to the North American plate. A significant portion of Haiti sits on the so-called Gonave microplate. The seismogenic source of the January 12 earthquake was near and believed at the time to be within the EPGFZ. However, further investigation has since indicated that the January 12 event was actually on a previously unmapped subparallel north-dipping fault, named the Léogâne fault (Calais et al. 2010, Prentice et al. 2010).

The relative movement between the Caribbean and North American plates is somewhat more complex than at some other plate boundaries. As seen in Figure 2-1, the mapped plate boundary, the Northern Hispaniola deformed belt, is located off the north coast of Hispaniola and is described as a subduction zone. In general, the North American plate is being obliquely subducted beneath the Caribbean plate as the Caribbean plate moves east-northeast (Dolan and Wald 1998). The slip rate at the plate boundary is approximately 19 mm/year. The east-west movement at this boundary is accommodated by subparallel left-lateral strike-slip fault zones within both the EPGFZ ($\approx$7 mm/year) and the SFZ ($\approx$8 mm/year) (Manaker et al. 2008).

Regional Seismicity

As with most plate boundaries, the Caribbean-North American boundary has a long history of seismic activity. The magnitude 8.1 event that generated a tsunami off the northern coast of the Dominican Republic in 1946 is a recent example. Since about 1500, seismic activity has generally alternated between the areas of the SFZ and the EPGFZ. Between 1615 and 1770, nine major events occurred near the EPGFZ while the SFZ was quiet. Then nine occurred near the SFZ and the Northern Hispaniola deformed belt between 1842 and 1953 while only one event occurred near the EPGFZ (Mann et al. 1998, Prepetit 2010). The last major event near Port-au-Prince occurred in 1770, and public awareness of the seismic risk was limited (Prepetit 2010). The estimated locations and relative magnitudes of historic earthquakes affecting the island are shown in Figure 2-2.

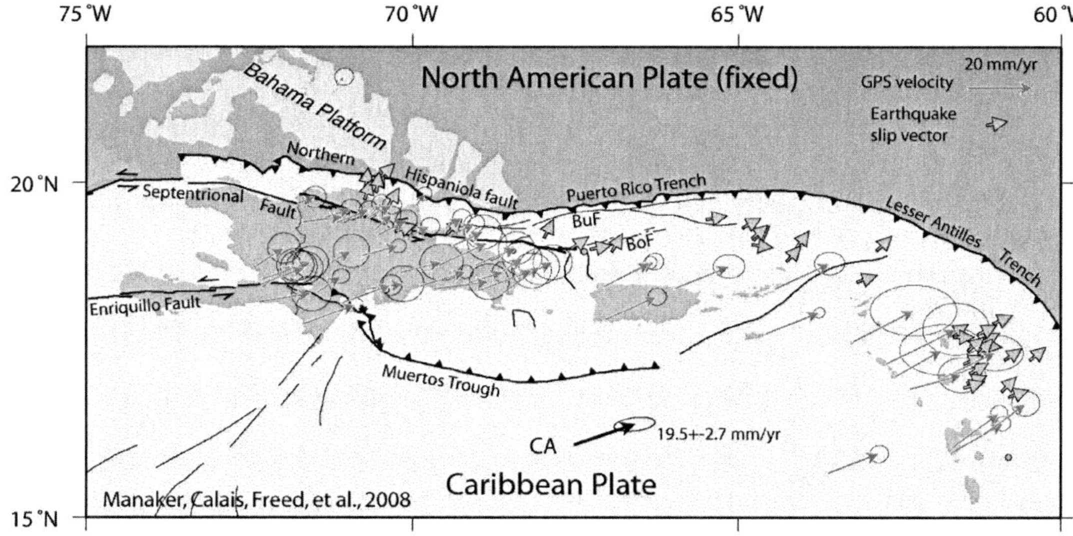

Figure 2-1. GPS velocities used to develop a kinematic model for the northeastern Caribbean (Source: Manaker et al. 2008, with permission from Eric Calais)

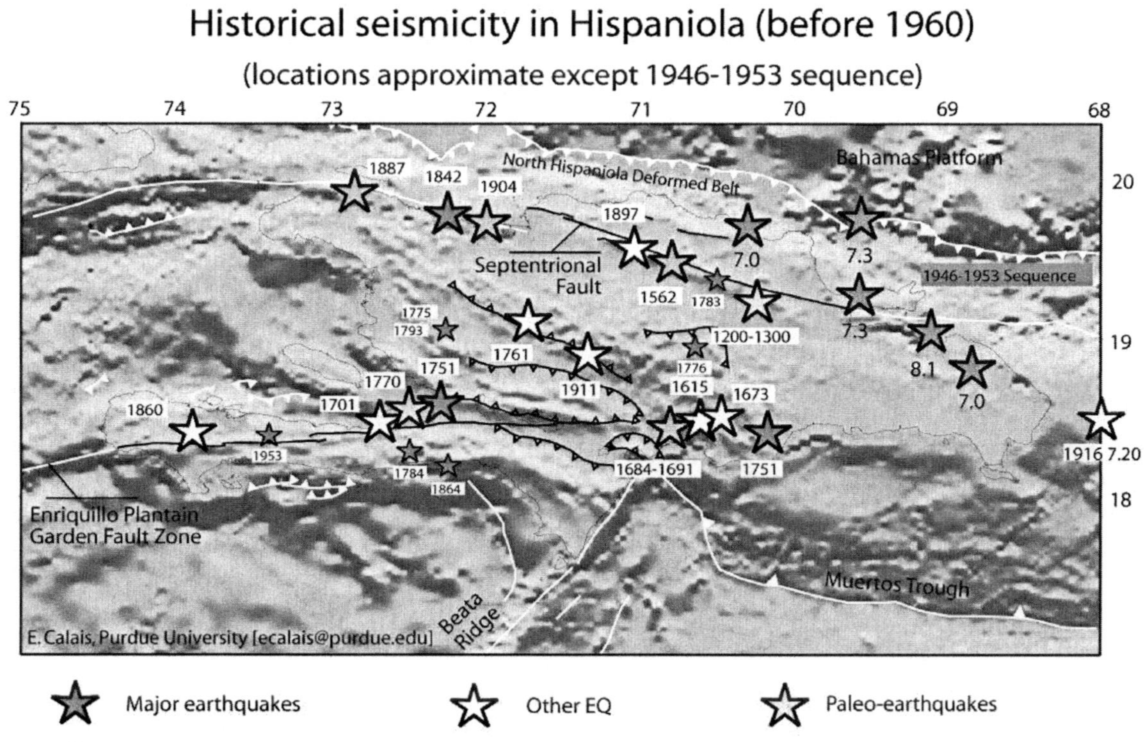

Figure 2-2. Approximate locations of historical earthquakes in the Hispaniola vicinity (Source: Purdue University Web site, with permission from Eric Calais)

In 2008 Manaker et al. published the results of a kinematic model of the plate motions in the northeastern Caribbean. Their model predicted that sufficient elastic strain was built up on the segments of the Enriquillo and Septentrional fault zones in Haiti to produce magnitude 7.2 and 6.9 earthquakes, respectively, if all the accumulated strain was released in a single event. As this strain does not appear to have been released in the January 12, 2010 event, the risk on the EPGFZ remains (Calais et al. 2010, Prentice et al. 2010).

January 12, 2010 Earthquake Characteristics

The January 12, 2010 earthquake occurred at 4:43:10 p.m. local time, was located at 18.457° N 72.533° W at a depth of 13 km, and had a moment magnitude (M_w) of 7.0 (USGS 2010). It was characterized by primarily left-lateral strike-slip movement propagating from east to west (Hayes 2010). No strong ground motion records were obtained for the main shock due to the lack of seismic monitoring stations within 100 km. Estimated distributions of the Modified Mercalli Intensity (MMI) and peak ground acceleration (PGA) as developed by the USGS are shown in Figures 2.3 and 2.4.

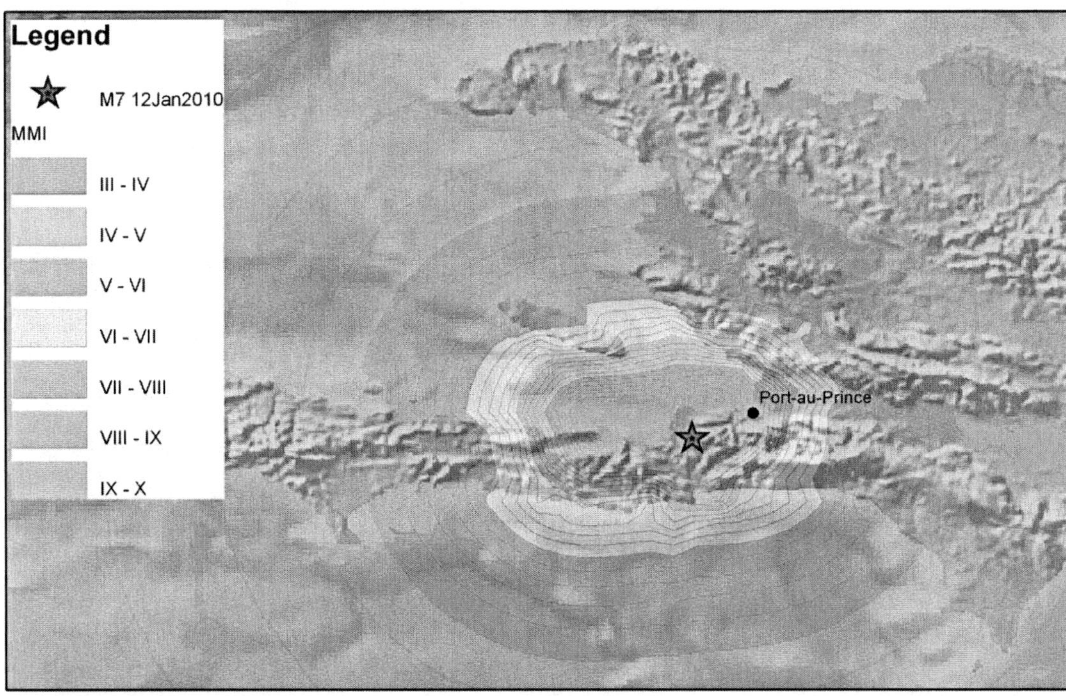

PERCEIVED SHAKING	Not felt	Weak	Light	Moderate	Strong	Very strong	Severe	Violent	Extreme
POTENTIAL DAMAGE	none	none	none	Very light	Light	Moderate	Moderate/Heavy	Heavy	Very Heavy
PEAK ACC.(%g)	<.17	.17-1.4	1.4-3.9	3.9-9.2	9.2-18	18-34	34-65	65-124	>124
PEAK VEL.(cm/s)	<0.1	0.1-1.1	1.1-3.4	3.4-8.1	8.1-16	16-31	31-60	60-116	>116
INSTRUMENTAL INTENSITY	I	II-III	IV	V	VI	VII	VIII	IX	X+

Figure 2-3. Shaded relief map of Haiti with location of the main shock and contours of the Modified Mercalli Intensity (MMI). (Source: USGS Shakemap Version 10 Shape Files)

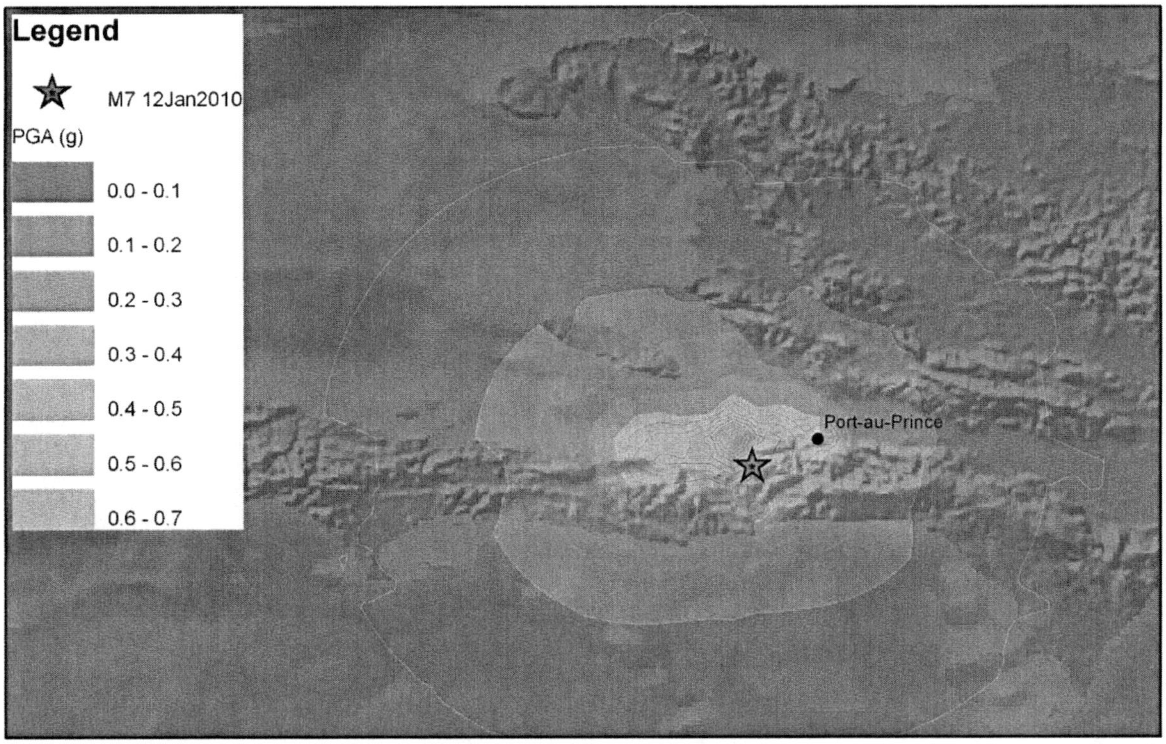

Figure 2-4. Shaded relief map of Haiti with location of the main shock and contours of the estimated peak ground acceleration (PGA). (Source: USGS Shakemap Version 10 Shape Files, accessed March 27, 2010)

The M_w 7.0 event was also characterized by a continuing series of aftershocks, with 59 events of M_w 4.5 or greater recorded through February 23, 2010. Most of the aftershocks had magnitudes between 4.0 and 5.0, as seen in Figure 2-5; however, smaller aftershocks have likely not been accounted for due to the lack of local monitoring stations. The main shock was followed by a M_w 6.0 and a M_w 5.7 within 20 minutes as well as a M_w 5.9 on January 20, 2010 (USGS 2010).

Coseismic Effects

Coastal uplift was observed in several areas and detailed investigations are in progress. Observed uplift was reported to be on the order of 0.5 m and extended for at least 8 km along the coastline between L'acul and Léogâne (Rathje et al. 2010).

Focused investigations by earlier teams did not reveal any evidence of surface faulting near the rupture zone (Eberhard et al. 2010, Rathje et al. 2010). These dominating factors prompted additional studies and led to additional research suggesting that a fault other than the EPGFZ was responsible for this event (Calais et al 2010, Prentice et al. 2010).

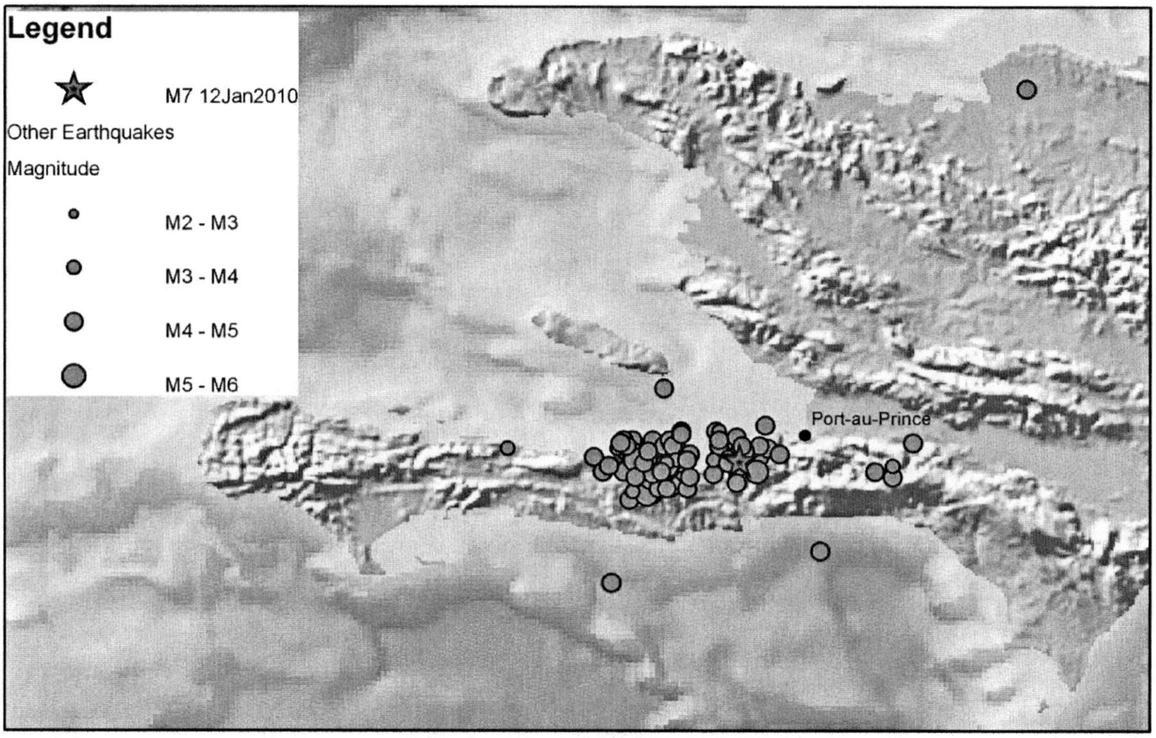

Figure 2-5. Shaded relief map of Haiti with locations of the main shock and following earthquakes from January 12 through March 28, 2010. (Source: USGS/NEIC (PDC) Data Base, accessed March 28, 2010.)

2.2 Geotechnical Issues

Liquefaction and Lateral Spreading

Observations of liquefaction and lateral spreading affecting lifelines were confined to the port of Port-au-Prince and are described in detail in Chapters 6 and 7. Other instances were reported by the team investigating land deformations along the coastline near Léogâne (Rathje et al. 2010).

Landslides and Rockfalls

Soil and rock debris was a hindrance to road travel in the weeks following the earthquake; workers were still actively clearing some segments of roadway during our investigation. The majority of failures occurred in steep, near-vertical road cuts made in hard soil/soft rock (primarily carbonate materials, such as limestone). Specific failures affecting roadways are discussed in Chapter 6.

Seismically-Induced Settlement

Several sections of roadway appear to have settled, causing cracking in the asphalt. However, without detailed soil investigations, the precise mechanisms for these failures cannot be determined. Photos and descriptions of representative failures are included in Chapter 6.

Ground Motion Amplification

Based on conversations with local residents and others investigating structural damage (personal communication with Anna Lang, March 14, 2010), it seems likely that ground motion amplification may have contributed to some of the damage; however, more detailed soil studies would need to be conducted and combined with information on geographical variations in building techniques before direct causation can be confirmed.

2.3 References

Calais, E., Freed, A., Mattioli, G., Amelung, F., Jónsson, S., Jansma, P., Hong, S.-H., Dixon, T., Prépetit, C., Momplaisir, R. (2010). "Transpressional rupture of an unmapped fault during the 2010 Haiti earthquake." *Nature Geoscience*, vol 3 no 11, 794-799.

Dolan, J. F. and Wald, D. J. (1998). "The 1943-1953 north-central Caribbean earthquakes: Active tectonic setting, seismic hazards, and implications for Caribbean-North America plate motions." *in* Dolan, J. F. and Mann, P. eds., "Active Strike-slip and Collisional Tectonics of the Northern Caribbean Plate Boundary Zone." *Geological Society of America Special Paper 326*, 143-170.

Eberhard, M. O., Baldridge, S., Marshall, J., Mooney, W., and Rix, G. J. (2010). "The *Mw* 7.0 Haiti Earthquake of January 12, 2010: USGS/EERI Advance Reconnaissance Team Report." *USGS Open-File Report* 2010-1048.

Hayes, G. (2010). "Finite Fault Model: Updated Result of the Jan 12, 2010 Mw 7.0 Haiti Earthquake." *USGS/NEIC*, <http://earthquake.usgs.gov/earthquakes/eqinthenews/2010/us2010rja6/finite_fault.php> (March 28, 2010).

Manaker, D. M., Calais, E., Freed, A. M., Ali, S. T., Przybylski, P., Mattioli, G., Jansma, P., Prepetit, C., and de Chabalier, J. B. (2008). "Interseismic Plate coupling and strain partitioning in the Northeastern Caribbean." *Geophys. J. Int.*, 174, 889-903.

Mann, P., Prentice, C. S., Burr, G., Pena, L. R., and Taylor, F. W. (1998). "Tectonic geomorphology and paleoseismology of the Septentrional fault system, Dominican Republic." *in* Dolan, J. F. and Mann, P. eds., "Active Strike-slip and Collisional Tectonics of the Northern Caribbean Plate Boundary Zone." *Geological Society of America Special Paper 326*, 63-123.

Mann, P., Taylor, F. W., Edwards, R. L., and Ku, T.-L. (1995). "Actively evolving microplate formation by oblique collision and sideways motion along strike-slip faults: An example from the northeastern Caribbean plate margin." *Tectonophysics*, 246, 1-69.

Prepetit, C. "Tremblements de Terre en Haiti: Mythe or Realite?" Laboratoire National du Batiment et des Travaux Publics d'Haiti, <http://www.lnbtp.gouv.ht/publications /index.html> (Mar. 19, 2010).

Prentice, C.S., Mann, P., Crone, A.J., Gold, R.D., Hudnut, K.W., Briggs, R.W., Koehler, R.D., Jean, P. (2010). "Seismic hazard of the Enriquillo-Plantain Garden fault in Haiti inferred from palaeoseismology." *Nature Geoscience*, Vol. 3 no 11, 789-793.

Purdue University Web site, http://web.ics.purdue.edu/~ecalais/haiti with permission from Eric Calais

Rathje, E., Bachhuber, J., Cox, B., French, J., Green, R., Olson, S., Rix, G., Wells, D., and Suncar, O. (2010). Geotechnical Engineering Reconnaissance of the 2010 Haiti Earthquake, Version 1.

U.S. Geological Survey (USGS). (2010). "Magnitude 7.0 – Haiti Region." <http://earthquake.usgs.gov/earthquakes/eqinthenews/2010/us2010rja6> (March 28, 2010).

USGS/NEIC (PDC) Data Base, http://earthquake.usgs.gov/earthquakes/eqarchives/ epic/epic_global.php, accessed March 28, 2010.

USGS Shakemap Version 10 Shape Files, <http://earthquake.usgs.gov/earthquakes/ shakemap/global/shake/2010rja6/#download, accessed March 27, 2010)

3. Water, Wastewater, and Drainage Systems

3.1 Description of Systems: Water Supply, Wastewater Disposal, and Drainage

Port-au-Prince Water Supply

There were three water supply systems or methods serving the residents of Port-au-Prince before the earthquake; an additional method was added after the event.

The first system is owned and operated by Centrale Autonome Metropolitaine d'Eau Potable (CAMEP) and is being replaced by Direction Nationale de l'Eau Potable and de l'Assainissement (DINEPA), the National Bureau of Potable Water and Sanitation. DINEPA was established in May 2009 and will reorganize the national water distribution operations. We met with the directors of each of these agencies and visited their facilities. The water delivery system serves Port-au-Prince as well as some of the smaller suburban communities.

The first system includes a supply and pipeline distribution system, which has approximately 250,000 connections serving more than 1 million people. There was no indication that services were metered. Water is plumbed into an unknown number of houses. Many houses have cisterns constructed on their roofs. These cisterns are filled by the CAMEP piped system and augmented by trucked water. In addition there are many public distribution kiosks throughout the service area where people can acquire water in their own containers. Before the earthquake there were rotating curtailments of water throughout the system, and areas would lose service on occasion due to inadequate supplies. The frequency of these curtailments is unknown.

The CAMEP supply system consists of 17 springs (collected in spring boxes, locally referred to as captages) and/or wells located around the periphery of the system at higher elevations in the surrounding hills. It appeared that most of the system was gravity fed except for the wells and one pump station serving a small area. Each of the systems has spring boxes and/or wells, chlorination facilities, and small storage tanks with capacities from 3,000 to 10,000 gallons. The tanks we observed were fiberglass. There are two privately-owned elevated concrete tanks in the system that were not in use at the time of the earthquake—one 40,000-gallon tank built in 1947 and the other more modern tank estimated to have 50 percent more capacity. CAMEP operates one elevated storage tank adjacent to the airport in Port-au-Prince but was unable to provide any details on it, other than it was not damaged by the earthquake. CAMEP and DINEPA indicated that the biggest issue with existing systems was the reliability and consistency of the liquid chlorination systems.

The system has one 3000 m^3 in-ground, cast-in-place concrete reservoir downstream. Transmission pipelines are constructed of ductile iron pipe with diameters up to 24 in. Distribution pipelines were PVC and galvanized iron pipe with diameters ranging from 2 to 4 in. According to CAMEP, there is a total of approximately 70 km of pipe in the system, which is a small length of pipe considering the number of connections. By comparison, a system serving the same number of connections in the United States might have 1000 km of pipe or more.

The water in this system is not suitable for drinking by U.S. standards without additional treatment. Most of the supplies had chlorination systems, but many were not operational before the earthquake due to lack of maintenance, unavailability of parts, and/or inadequate supply of chlorine. Water quality tests from the distribution system performed by both the U.S. Air Force and a German emergency response organization showed positive coliform counts and high hetero plate counts (HPCs) in the water.

There are ongoing attempts to limit surface water drainage into the spring boxes to maintain water quality. At the facilities we visited, it appeared there was potential for contamination of the collected spring water.

CAMEP is trying to expand the system to serve more people. They have seven wells in Pernier that are not used because no pumps, piping, or treatment facilities have been installed.

CAMEP has five administrative buildings throughout the system including office space and customer service facilities where people pay their bills.

The second system is a privately-owned network of wells and groundwater supplies that are treated and distributed by tank trucks. If the public CAMEP system was unable to fill a cistern, residents could buy water from one of the tank trucks that circulated through the area playing music to notify people they are coming. The groundwater supply for these systems appears similar to the public system of springs and wells. Treatment facilities for this second system are in place using technologies such as filtration, reverse osmosis, and ultra violet and chlorination disinfection, but it appeared that they were not fully functional before the earthquake. This second system was not being fully used after the earthquake because water was being provided free of charge by the various international agencies assisting with the earthquake recovery. Thus, people who had purchased water from this second system before the earthquake could now get it free from the CAMEP system, which was being treated by various portable treatment facilities. As a result, it was difficult for the owners of this private system to sell water.

The third system consists of water sold in 5 gallon bottles (similar to those found in offices in the United States) distributed at kiosks throughout the area at a cost of $1.50 each. This system was not being fully used following the earthquake because of the free water provided by the various international aid agencies.

A fourth system, provided by various international governments and NGOs, was implemented following the earthquake. A series of small portable water treatment units using various water treatment technologies were set up to treat the public water supply system for potable consumption. Water from these units was collected locally by residents in plastic containers. In some locations, water from these facilities was pumped into tank trucks and distributed to tanks and bladders that had been distributed around the region.

Jacmel Water System

The water system in Jacmel, a small city on the southern coast of Haiti, is owned and operated by Service Nationale d'Eau Potable (SNEP). SNEP, which is also being replaced by DINEPA, is responsible for water supply in communities outside Port-au-Prince. The Jacmel system serves about 3000 customers (1500 to 1600 meter connections). The areas served include the Lamandou, Lopital, and Raquette zones . Sixty percent of customers are active, and 40 percent are inactive for nonpayment of their bills. It is unclear whether this was a pre- or post-earthquake status and whether it resulted in termination of services.

The water systems in Jacmel, similar to Port-au-Prince, are a combination of public augmented by private supplies. The public water supply consists of a series of spring water collection boxes. The site visited by the team had a 600 m^3 cast-in-place concrete reservoir, which was fitted with chlorine disinfection facilities using HTH (calcium hypochlorite). The water is then transmitted into town through a cast iron line with leaded joints built in 1946-47. The chlorine system was not operable before the earthquake. The Jacmel system was taken over by SNEP in 1984, at which time some rehabilitation was performed.

The distribution system is reported to have 4 to 5 km of 4-in. schedule 40 PVC pipe with solvent welded joints. SNEP does not have the design drawings.

The Spanish are promoting the use of high-density polyethylene (HDPE) pipe with fused joints for new systems in rural areas. This pipe is similar to that used by many gas utilities in the United States.

Port-au-Prince Wastewater Disposal and Drainage Systems

The wastewater disposal systems include septic tanks with leach fields, cesspools, private latrines, and direct discharge of raw sewage to the storm drainage system. Unofficial estimates based on discussions with NGO representatives and field observations are that somewhere between 10 and 30 percent of the area is served by septic tanks and another one-third by cesspools or latrines, with the balance discharging to the stormwater drainage system.

The only wastewater treatment plant (WWTP) in Port-au-Prince is at the National Hospital, and the plant has not been in operation for several years. Wastewater is presently bypassing the plant and draining into the ocean.

Large, modern buildings have septic systems. In discussions with owners, it became apparent that the septic systems required significantly more maintenance than those we are familiar with in the United States.

The stormwater drainage systems consisted of deep, open channels beside the roads, which discharge into large, open channels that collect stormwater from the mountains and impervious areas (Fig. 3-1). In some locations, the roadside channels were covered with concrete panels.

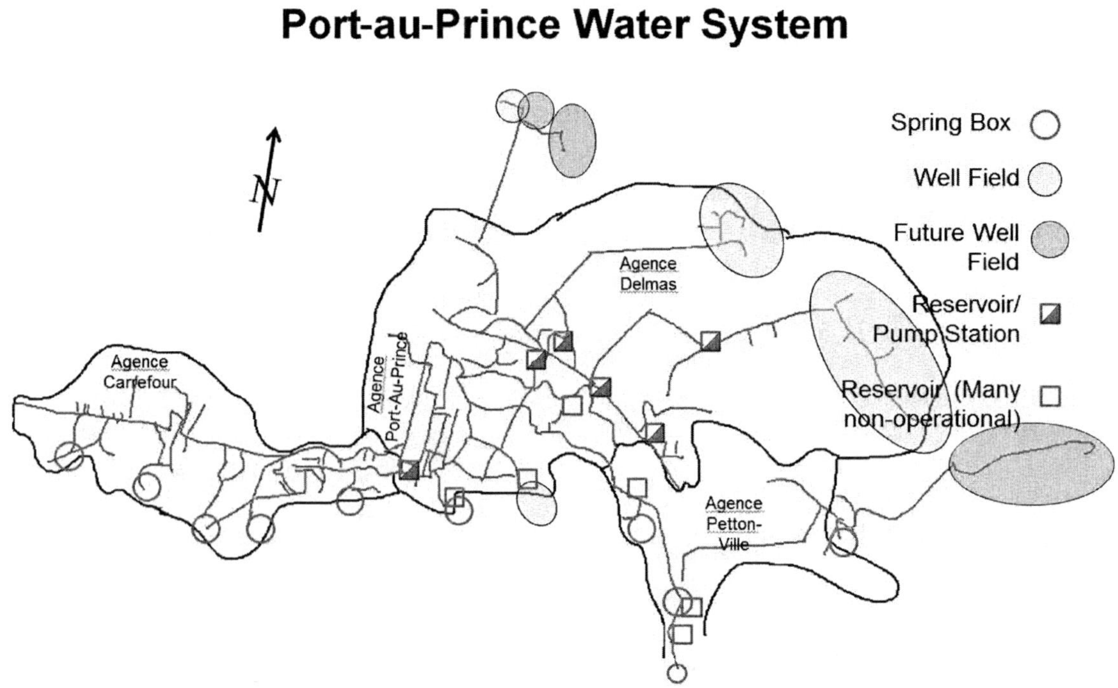

Figure 3-1. Priority channels in the Port-au-Prince drainage system

3.2 Earthquake Performance

Port-au-Prince Water System

CAMEP reported that they lost five employees in the earthquake. In addition, the earthquake caused a loss of 50 percent of paying customers due to loss of life and/or loss of home or business. As a result, they have experienced inadequate revenue to cover payroll.

The public water system suffered only minor damage to most facilities. The day following the earthquake, most of the pipeline breaks were isolated. It took less than a week to restore service. Because all water mains may not be in service, additional unidentified pipeline damage may become apparent as lines are eventually returned to service. CAMEP reported eight to 10 pipeline breaks in 70 km of pipe. This is a very low break rate considering the extent of other types of damage that occurred.

We visited the Turgeau water collection facility, which has a spring box, a well, and small fiberglass storage tanks (Fig. 3-2) (N18 31.549 W72 19.250). There appeared to be no damage. Action Contre la Faim, ACF International Network (a French NGO) had set up a small temporary treatment plant using the CAMEP groundwater supply (Fig. 3-3). The water was being distributed through a header with spigots (Fig. 3-4) and by tank trucks filling up for delivery to tanks throughout the area. Just downstream

Figure 3-2. Turgeu fiberglass storage tanks

of the Turgeau captage, there was an undamaged cast-in-place 1180 m^3 concrete tank (Fig. 3.5). There was extensive residential building surrounding the CAMEP facility (Fig. 3-6). Adjacent to the ground storage tank, the Université Quisqueya campus was heavily damaged, and it was reported that 50 fatalities occurred in a collapsed building. In a ravine below the campus, a major water transmission main supported on columns failed due to a landslide. This pipeline was repaired with what appeared to be a solid sleeve pipe clamp.

One pipeline failure occurred in Bourdon, where a 300-m section of pipe was damaged. The details are unknown as it was repaired before our visit. The cast-in-place concrete reservoir at Bourdon had some interior damage (N18 32.207 W72 18.842). Additionally, there were many service line failures, primarily in PVC pipe and particularly in the Bellevue area, where there were many collapsed buildings. Any operable chlorination systems were being powered with emergency generators. No information on the number, locations, and duration of operation was provided.

The biggest issue following the earthquake was getting potable water to the estimated 1 million people displaced. Port-au-Prince residents could not stay in their houses because they had collapsed or could potentially collapse. Many of the refugees were staying in temporary tent camps located throughout the city. At the time of our visit, foreign emergency response organizations and NGOs had set up portable treatment equipment at selected locations and were treating water drawn from the CAMEP public water system. Organizations from the United States, Germany, France, and Mexico were identified. The largest treatment facility visited was provided and operated by an emergency response unit from Germany. This major temporary treatment facility was located on the CAMEP campus near the airport (Fig. 3-7, 3.8), and relief workers were distributing potable water with tank trucks supplied by local contractors. On January 19, water began to be distributed to 500 sites that had temporary plastic tanks and bladders.

Figure 3-3. Turgeau emergency treatment plant

Figure 3-4. Water distribution at Turgeau

Figure 3-5. Undamaged concrete reservoir downstream from Turgeau

Figure 3-6. Collapsed house next to concrete reservoir in Figure 3-5

Figure 3-7. German water treatment plant near airport

Figure 3-8. German water treatment plant

Figure 3-9. Laboule facility sign

We observed a privately-owned water supply and treatment facility at Laboule 12 (Fig. 3-9) (N18 29.777 W72 19.761). It pumped water from a well up to a cast-in-place concrete storage tank approximately 10 m by 10 m by 5 m deep. Treatment was provided by a reverse osmosis unit that was not operating (Fig. 3-10). Power was provided by a generator, as power from the state utility, Electricité d'Haiti, was sporadic even before the earthquake. Water was then distributed by tank truck (Fig. 3-11) or by selling it to walk up residents who provided their own containers.

Water was being supplied to the camp at the national soccer stadium by a portable water treatment plant provided by Operation Blessing International from Virginia and Water Missions International from South Carolina. Inside, 800 families were set up in tents with another 700 families in the adjacent parking lot (Fig. 3-12). The system was pumping water from the public water system and treating up to 10,000 gallons per day (Fig. 3-13).

At the National Hospital, a Spanish emergency response organization had set up a small emergency water treatment facility, pumping water from the main underground storage reservoir, treating it, and distributing for use at the hospital. Nearby, there was a solar power heater setup to provide a hot shower.

Two private concrete elevated tanks are located within the National Hospital compound. Neither had been in operation before or during the earthquake, and they remained out of service. Both tanks showed signs of distress at the joints from earthquake ground motion (Fig. 3-14, 3.15, and 3.16).

Figure 3-10. Laboule water treatment plant

Figure 3-11. Truck loading water at the Laboule WTP

Figure 3-12. National soccer stadium

Figure 3-13. Water distribution outside the national soccer stadium

Figure 3-14. A 40,000 gallon elevated tank at National Hospital

Figure 3-15. Newer elevated tank at National Hospital

Figure 3-16. Distress at joint in newer elevated tank at the National Hospital

DINEPA was in discussions with the National Coalition of Trucks to reestablish delivery of water from the second system, privately-owned treated water. The intent is to reestablish regular (pre-earthquake) water service when people return to their houses, which was estimated to be several months away.

SNEP Water Systems Including Jacmel

The most significant damage occurred to the facilities in Leogane, Grand Goâve, Petit Goâve, and Jacmel. Some damage also occurred to systems in Gressier, Léogane, Grand Goâve, Petit Goâve, Miragoane, Anse-à-Veau, Croix-des-Bouquets, Jacmel, Thomazeau, Arcahaie, and Cabaret. In general the system suffered when buildings collapsed, severing the service lines.

The water system in Jacmel was partially operational. The cast iron transmission piping from the main water sources was intact (Fig. 3-17). Of an estimated half dozen spring boxes in the Jacmel system, one was severely cracked when a rock outcropping above it shifted (Fig. 3-18 and 3.19) (N18 15.751 W72 29.576). Downstream at the concrete reservoir (Fig. 3-20), a 12-in. gate valve had been replaced with the assistance of the Canadian military (Fig. 3-21) (N18 15.712 W72 29.569).

Figure 3-17. Intact cast iron pipe at Jacmel water source

Figure 3-18. Cracked spring collection box at Jacmel

Figure 3-19. Close-up of cracked spring collection box

Figure 3-20. Reservoir downstream

Figure 3-21. Valve replacement at Jacmel reservoir

SNEP reported many leaks in the distribution system (presumably dozens); however, the Canadian military personnel helping them restore the system said that they had not found any pipe failures. SNEP was in the process of replacing system valves, many of which would not hold, and reported that 4-in. repair clamps were in short supply.

Figure 3-22 shows a typical bladder at a camp used for drinking water distribution in Jacmel.

The Petit Goâve system apparently had several leaks. The spring box had been repaired, but the 300 m^3 reservoir was cracked. It was not reported whether the cracks were critical. The water system in Leogane was not operational before the earthquake.

Figure 3-22. Bladder used for water distribution at camp in Jacmel

Wastewater and Drainage

In general, the earthquake had little direct impact on the drainage system. In a limited number of situations, there was damage to the facilities from landslides, collapse of embankments, and differential settlement. In the lower lying areas, there was sediment and debris buildup that blocked the channels. It appeared most of the material had been there before the earthquake. In some cases, buildings collapsed into drainage channels causing further blockage. There were reports of septic systems not working as a result of differential settlement of connecting pipelines.

The only identified wastewater treatment plant in Port-au-Prince, the one at National Hospital (Fig. 3-23), had no apparent damage. Damage can sometimes be identified because a facility becomes inoperable; however, damage is more difficult to identity in this case because the facility was inoperable before the event. A team member had observed the treatment facility several months before the earthquake and reported it was already inoperable at that time. On January 12, work was under way to drain the WWTP basins to determine what could be done to return it to operation. This work was initiated by Operation Blessing International from Virginia.

The two most significant issues following the earthquake were (1) how to deal with the waste generated by the large displaced population and (2) how to mitigate flooding in the low-lying areas where temporary camps had been set up.

Figure 3-23. Wastewater treatment plant at National Hospital

Latrines were constructed in most of the camps. Figures 3.24 and 3.25, respectively, show the overview of a latrine and details of its construction. In the short term, these seemed to provide adequate service. Figure 3-26 shows a technology being explored to separate the liquid and solid streams in latrines for easier disposal. As of early March, there were plans to deploy the first of this type of facility. In some cases, portable toilets had been deployed. The issue remains of what to do with the waste. A similar issue is disposal of septage (sludge) from septic tanks. It was being dumped in an unregulated landfill and in some cases on an open field north of Port-au-Prince, both creating a potential for human contact with the waste and groundwater contamination.

In recent years, drainage channels have not been well maintained. Water carrying debris, sanitary waste, and sediment drains from the steep watersheds in the mountains surrounding Port-au-Prince and the impervious areas in the city. As the gradient flattens and the water reaches the tidal backwater from the ocean, the velocity slows and allows the solids to settle in the channel. As a result, the drainage channels fill up reducing their capacity. During wet weather periods, there is potential for flooding. As the drainage system carries a significant amount of the human waste, the floodwater is contaminated. The rainy season in Haiti starts as early as March. The hurricane season starts in June, compounding the problem.

Figure 3-24. Latrine at camp in Jacmel

Figure 3-25. Details of latrine in Jacmel

Figure 3-26. Toilet seat designed to separate liquid and solid waste at the source

Before the earthquake and when the drainage channels were maintained, the contaminated drainage/waste was washed out into the ocean. In more recent years, flooding occurred as the channels filled with debris. Because of the earthquake, hundreds of thousands of people are now in camps subject to flooding, and exposure to contaminated floodwater has dramatically increased.

DINEPA had developed a plan to clear the downstream segments of the most critical rectangular drainage channels. Table 3.1 summarizes the channels in order of priority. As of early March 2010, DINEPA was hopeful of initiating a program by the end of the month that would allow them to quickly clear these channels.

3.3 Restoration and Future Plans

DINEPA is planning to decentralize the water system, making each municipality responsible for its own supply. Its three-year goal is to increase the number of cities with clean potable water. A new 2009 law allows the privatization of water systems, which is being considered in some locations. Privately-owned water systems have been started in four or five cities, including St. Marc. Some of these may be public/private partnerships where DINEPA will administer the contracts. International Development Board (IDB) and the Spanish are investing in various cities. Rates will be set depending on location and social and economic situation.

DINEPA is also evaluating regulations for private sanitation.

Table 3.1. Drainage Channels Requiring Debris Removal to Mitigate Flooding

Channel Name	Start Point	End Point	L (km)	W (m)	D (m)	Waste/ Debris Vol. (m3)
Georges	Rue Colza	Coast	2.81	10.0 to 20.0	2.5	70,200
Bois de Chene	Ave Christophe	Coast	3.08	10.0 to 20.0	2.5	77,200
Orphelin	Route de Delmas	Coast	3.80	5.00 to10.0	2.5	20,000
Delmas	Ruelle	Coast	5.21	5.0 to 20.0	2.5	72,500
Eh	300 m E of Rte Nat'l	Coast	1.72	10.0 to 20.0	2.5	43,800
Brea	Rte Nat'l # 2	Bois de Chene	0.67	5.0 to 10.0	2.5	10,100
Maleli	Blvd Dessalines	Bois de Chene	0.94	5.0 To 10.0	2.5	14,100
Papacito	100 m S of Rte Nat'l 2	Coast	0.51	5.0 to 10.0	2.5	7,700
Madam Gano		Coast	1.52	5.0 To 10.0	2.5	22,800
TOTAL			20.26			338,400

Source: *Data provided by DINEPA*

Reconstruction in Haiti could consider planned residential/commercial/market areas on the Northeastern side of Port-au-Prince. This area appears ready for development and with proper planning could provide housing and jobs for displaced persons and model water, wastewater, and drainage systems to begin to improve the quality of life in Port-au-Prince. With a planned development, residents from central Port-au-Prince could be offered the opportunity to relocate to this area, which would simplify the rebuilding of the urban core.

3.4 Major Observations and Recommendations

Four major observations were made:

1. The water and drainage systems were in poor condition before the earthquake, which had only limited impact on these two systems.
2. Both the water supply and drainage systems had limited infrastructure (e.g., few kilometers of pipe, pump stations, etc.) compared to comparable U.S. communities serving the same size population.
3. The biggest issue confronting the region was providing water and waste disposal services to the large displaced population.
4. During the TCLEE team investigation, emergency water supply was being adequately handled by foreign emergency response teams. The issue was maintaining that service once the emergency response teams left.

Three recommendations are offered:

1. Water and drainage systems should require a minimum level of operation and maintenance.
2. A program should be developed to provide ongoing training to design, construct, operate, and maintain water and drainage facilities.
3. Use of HDPE pipe is recommended for use in new distribution facilities because it is highly resistant to earthquakes.

3.5 Acknowledgments

The ASCE TCLEE water and drainage investigation included Don Ballantyne, Dave Plum, and Brucely Joseph. Field reconnaissance was conducted from Monday, March 1, through Friday, March 5, 2010, about 7 weeks after the event.

The TCLEE team would like to acknowledge the following people in Haiti who provided information for the report:

- Patrice Roy, Director of CAMEP (Centrale Autonome Metropolitaine d'Eau Potable) Ing. Central Bureau in Port-Au-Prince
- Pierre Philippe Jean, Director of SNEP (Service Nationale d'Eau Potable), Delmas 27 Bureau
- Esterne Michel, SNEP Southeast Regional Director (Directeur Departmental du Sud-Est) Ing.
- Gerald Jean-Baptiste, Directeur General de DINEPA (Direction Nationale de l'Eau Potable and de l'Assainissement) National Bureau of Potable Water and Sanitation
- Wilston W. Etienne, P.E., Principal and Program Leader, Golder Construction Services, Atlanta, GA, USA

4. Energy Sector

4.1. Description of System

This chapter reviews the infrastructure damage to the government-owned Electricité d' Haiti's (EDH) system and its service area within and near the Port-au-Prince (PAP) metro area. EDH is a government owned and operated bureau under the jurisdiction of the Ministry of Public Works, Transportation, and Communications. EDH's creation in 1971 followed the construction of the electric generation unit and powerhouse at the Peligre Dam and Reservoir. The latter was constructed in the early 1950s for flood control and water supply to the region.

Peligre's hydrogeneration is augmented by thermal generation plants located in the Delmas (67.95Mw) and Carrefour districts. The Carrefour generation complex comprises two distinct facilities: Carrefour-1(49.92Mw) is EDH owned and operated, along with its associated transmission, feeder, and distribution systems. Carrefour-2 is owned by the Venezuelan government and operated by a South Korean concern. The construct of the distribution system ranges from a standard similar to some U.S. systems with a multi-grounded neutral (MGN) under-built of the primary conductor(s) to systems where grounding is attained at the end user's location. Other unique build-outs occur at commercial locations where the business or property owners construct their own pole, step-down transformer, and service drop at the end use location. According to our sources, they assert that these types of installations interconnecting with the grid are completed without the knowledge or authorization of EDH management. We noticed several other locations with similar installations.

Additionally, we saw private conductors plugged into the power side of vacant meter bases or in some case tapped (wire-wrapped around an insulation-stripped barred conductor) onto one conductor of a service secondary supply side. These two factors corroborate the assertion of Myk Manon of the National Rural Electric Cooperative Association (NRECA) that 25 percent of the EDH system is meter monitored while only 40 percent of these users reimburse the operator for their accumulated usage. According to the revised National Energy Plan, EDH's objective, set in 2006, is to increase both the number of metered services and the number of clients paying their monthly bills on a recurring basis over the next 15 years. This objective is one of many intended to make the EDH a commercially viable operation.

4.2. Overview of System Performance

Generation-Peligre/ Hydro

The Peligre system is capable of producing 54Mw. According to EDH's onsite representative, the normal power output estimate is approximately 7.5 to 9 percent of the facility's capability, due in part to severe reservoir sedimentation, drought, and aging generating equipment, which is subject to frequent breakdown and lack of spare parts. This facility was not damaged during the January 12, 2010 earthquake; however, Peligre halted power generation for a short period to isolate the system from downstream facilities to ensure a safe work environment for the initial response efforts in Port-au-Prince.

Generation-Carrefour-1/Thermal

Seismic impact to the thermal generator will focus on EDH's Carrefour-1 (49.92Mw capability), which sustained significant structural damage to the building housing the seven diesel generators, including noticeable settlement on the grounds west of the building (Fig. 4-1). Secondly, the oil cooling system suffered a pipeline breach (Fig. 4-2) and resulting oil spill. Further damage included the dislodging of a rail-mounted transformer (Fig. 4-3), which resulted from the vertical movement of the transformer during seismic activity and the lack of a mechanical restraint. Figure 4.4 illustrates that horizontal (lateral) movement was constrained by two metal cleats at the juncture of the transformer's casters and the rails mounted on concrete piers. The cleats are secured in place by U-bolts, which measure about .95 cm in diameter. This is a practice that should be standardized throughout the system, Currently, EDH has ceased its operations at Carrefour-1 until it decides whether to rebuild or construct a new facility at this or another location.

Our visit to the Delmas substation reveals inadequacies in the installation of safeguards for rail-mount transformers, installed conceivably relying on vertical dead weight friction to keep the equipment in place. The following illustrations (Fig. 4-5 to 4-8) detail the vertical (25.4 cm) and lateral (243.04 cm) movements of a transformer that jumped off its mount, travelling to its resting place and then dislodging, but not damaging, an adjacent oil-coolant reservoir tank. The position shift damaged the transformer's oil cooling reservoir petcock and severed the rigid buss work.

Additionally, a similarly installed adjacent transformer broke loose from its mount and severed its rigid buss work. Response efforts and system functionality was temporarily restored, aided by Dominican Republic's Power Authority crews with the installation of five trailer-mounted, diesel-powered generation stations, implemented to carry the downstream load. This allowed EDH and subcontract crews to isolate and restore the transformers on temporary steel I-beam mounts. We arrived to see the transfer of load taking place, thus ending the need for a portable generator. Unrelated to the damage sustained at Delmas, the quality of workmanship is as important to system resilience as proactively installed seismic safeguards. Figure 4-9 shows the splice of hand-twisted conductors with minimal clamping across the work.

Figure 4-1. Building damage at Carrefour-1

Figure 4-2. Breach in oil cooling pipeline at Carrefour-1

Figure 4-3. Dislodged transformer at Carrefour-1 caused by vertical movement

Figure 4-4. Transformer caster cleats, attempt to minimize equipment lateral movement at Carrefour-1

Feeder and Distribution Overhead Systems Overview

The TCLEE team viewed the impairment sector's feeder and distribution systems in the PAP metro area, where service delivery operations were impaired and/or facilities were destroyed due to the collapse of numerous buildings in close proximity to overhead infrastructure. The support structures for EDH's overhead utility systems are made of either wood or reinforced concrete poles. Further investigation revealed that wood support structures and their transmission tower counterparts had the flexibility to sustain the seismic impacts, except for those situations where buildings collapsed on overhead facilities. EDH's construction policy is to use smaller gauge tie lines in some cases at insulator weak points as sacrificial elements to maintain lines and support structure integrity, thus creating a loss of service rather than a loss to a capital asset (see Figure 4-11). Our team noticed that more support structure failure occurred with concrete poles (see Fig. 4.12 to 4.13).

Figure 4-5. NRECA's Myk Manon stands where the Delmas substation transformer, 91.44 cm to his left, jumped off its mounts, landing and damaging the reservoir tank and standpipe on his right.

Figure 4-6. A broken transformer caster and damaged concrete pier at the Delmas substation

Figure 4-7. Rigid buss work replaced with flexible connections the Delmas substation

Figure 4-8. Former rigid buss work the Delmas substation

Figure 4-9. Hand-twisted neutral bond splice, minimal clamping at the Delmas substation

Figure 4-10. Transmission lines coming from the Peligre hydropower facility with a transmission system support composed of galvanized steel truss work

Figure 4-11. Broken tie wires

Figure 4-12. Broken concrete Pole-1

Figure 4-13. Broken concrete pole-2

Figures 4-15 to 4-17 illustrate some of the damage to overhead lines caused by the structural failure of the building immediately adjacent to the joint-use facilities (see Fig. 4.14). The over-built balcony of the building to the left of the pole (Fig. 4-15) creates not only a potential threat to the lines and supporting structure but also places the building occupants in very close proximity to primary power conductors. Damage to overhead facilities from the collapsed building occurs in the next span to the right (Fig. 4-15).

Operators should consider undergrounding overhead lines to mitigate service outages caused by building structures collapsing on overhead lines in urban environments, especially where zero lot-line setbacks are the norm. In underground construction, we suggest using slip connectors within the associated vault and conduit systems as a countermeasure to potential lateral movement, settlement, or subsidence, coupled with slack loops formed into supply cables, as the facilities are racked in vaults and at end-use locations. As an added benefit, an undergrounded urban system augments system security, provides added protection from accidental contact with energized equipment or conductors, and potentially minimizes illegal or bootleg connections, especially when the access lids are fitted with a mechanical locking system with a specially configured socket key.

Figure 4-14. Joint-use construct; note the street light secured to the pole with rope

Figure 4-15. Balcony overbuild above telecom space (Courtesy of Eduardo Fierro, BFP Engineers, Inc.)

Figures 4.16 (left) and 4.17 (right). Damaged joint-use facilities, from collapsing buildings (Courtesy of Eduardo Fierro, BFP Engineers, Inc.)

4.3. Recommendations

The following suggestions are provided to improve system performance. EDH has already embarked on a GIS-based mapping initiative and should consider continuing this project until it has a complete inventory of all transmission and distribution facilities by type, capacity, and approximate mortality date, including all distribution EDH transformers and those privately held. The following suggestions could be elements of a pilot project that, if proven successful and beneficial to the community, could be repeated in other service areas, as necessary:

- Provide an infusion of capital to upgrade or install control systems; determine and adopt appropriate SCADA-type systems.
- Rehabilitate, reconstruct, or convert existing transmission, feeder, and distribution systems to an acknowledged construction and operational standard to establish and accomplish routine maintenance across the entire system, thus, improving service delivery, energy assurance, and reliability.
- Invigorate existing workforce skill sets, as standards are adopted, with tools and methodologies to enhance system reliability, service delivery assurance, and workplace safety.
- Develop an apprenticeship program that trains and mentors new industry entrants in all technical occupational work groups—generation, line, and substation technical specialists.
- Re-allocate the system into hub and spoke segments within the Port-au-Prince core area—area substations (hub), including feeders and distribution systems (spokes)—and correlate those facilities and the estimated demand load to the neighborhood demographic identifying commercial, industrial, and residential enclaves.
- Consider distributed generation alternatives using renewable energy resources, such as bio mass, solid waste, wind, and solar, and assigning those generation resources to the identified demographic centers (districts) listed in Section 4.2.
- Encourage end-user generation by providing capital credits for the installation of renewable generation alternatives at commercial or industrial locations, including tax credits for property owners of single- and multi-family residences.
- Develop, design, and reconstruct boulevards and streets in the Port-au-Prince core area without any overhead facilities by undergrounding all electrical facilities employing voltages below 35 kV and ensuring that the resulting excavations have the capacity to allow multiple systems to occupy the trench-work, as tenancy groups are identified.
- Empower EDH with condemnation authority as necessary to establish, protect, and maintain its rights of way (ROW) and the authority seize privately installed electrical components and equipment illegally attached to the EDH system. For safety purposes, the ROW should be cleared of all buildings and structures for all system components above 65Kv, including all sub-transmission and transmission elements.

4.4 Acknowledgments

We would like to express our appreciation for the assistance from NRECA's Myk Manon in providing an overview of the EDH system and insight into the operational challenges the electrical system must overcome. We would like to acknowledge Eduardo Fierro and Cynthia Perry, BFP Engineers Inc., who granted use of their photographs. Additional gratitude is extended to Ms. Florence Grandiot and her provision of pleasurable accommodations, meals, and the needed rest from our days reviewing the devastation wrought by this disaster.

I would like to express my appreciation to Ms. Yumei Wang, P.E., of Oregon's Department of Geology and Mineral Industries, and Mr. J.R. Gonzalez, P.E., Administrator-Safety, Reliability, and Security Division of the Public Utility Commission of Oregon, for providing me the opportunity to participate in this endeavor. I came into this project knowing that I would gain more from this experience than the expertise I could contribute.

5. Telecommunications Sector

5.1 Description of System

The telecommunication systems reviewed are Teleco, a government-owned, single wireline carrier operated under the Ministry of Public Works, and three wireless mobile vendors using Global System for Mobile Telecommunication (GSM)[1] and code-division multiple access (CDMA)[2] formats. For wireless broadband service, a single Wi-Fi operator is available. From information provided through the U.S. Department of Homeland Security National Communication System (DHS-NCS) representative, there are four Internet service providers (ISPs) active in-country. This chapter focuses on Teleco and the mobile wireless operator Digicel.

Team members initiated several attempts (pre- and post-investigation) to contact wireline carrier representatives but did not receive a response or acknowledgement. As a result, the review team was at a disadvantage in attaining a complete analysis. Situational awareness and utility insight would have been valuable in understanding system response and recovery. Our system review is based on the team's field observations, expertise, and experience.

Teleco is a wireline-based utility provisioning service, similar to networks throughout the United States. A more descriptive analogy would be a hub and spoke matrix with a centralized tandem switch that interconnects with other tandems to exchange information (voice and data). End offices are subordinate to the tandem and only control the office's traffic within their specific wire center boundaries. The feeder and distribution outside plant (OSP) facilities are constructed in an aerial environment, much of which was supported on either concrete or wood poles, and are commonly placed in a joint-use configuration, which included coaxial cable (CATV) and power line constructs.

Digicel's system is comprised of antennas constructed on hilltops and buildings throughout the service territory. According to the U.S. Federal Communications Commission (FCC) assessment team, led by International Branch Chief Mindel De La Torre, Digicel's international traffic is transmitted out of country via micro-wave radio-frequency (RF) connections through the Dominican Republic to its destination. As such

---

[1] Global System for Mobile Telecommunication (GSM) comprises the CEPT-defined standardization of the services, functional/subsystem interfaces, and\protocol architecture, based on the use of worldwide standards produced by CCITT and CCIR, for a pan-European digital land mobile system primarily intended to serve users in motor vehicles. (IEE Communications Society Web site)

[2] Code-Division Multiple Access, a digital cellular technology, which utilizes spread-spectrum techniques, unlike GSM, CDMA utilizes every channel in the available spectrum. Individual conversations encoded with a pseudo-random digital sequence. CDMA consistently provides better capacity for voice and data communications than other commercial mobile technologies, allowing more subscribers to connect at any given time, and it is the common platform on which 3G is based. (Webopedia Web site)

the system is independent of the publicly switched telephone network (PSTN) operated by Teleco.

5.2 PSTN Wireline Telecommunications

The January 2010 seismic activity is reported to have caused the collapse of the Teleco building and isolated the end offices subordinate to the access tandem switch (ATS) and any outside plant (OSP) fed from this facility. Local service within the domain of end offices could still provide intra-office communication as long as commercial power, back-up generation, and battery reserves remained capable and intact with functional OSP. At several locations throughout Port-au-Prince and outlying communities we observed the placement of mobile cellular system and telescoping antenna array, also known as cells on wheels (COWS), outside of several telephone central offices, presumably placed as a work around to enable inter-exchange traffic.

In addition to impairing inter-office operations, the international call-traffic stopped immediately with the building's collapse.

Community Impact

Unlike the consumer-dependent U.S. systems, Haiti's PSTN serves a very small segment of the population, business, and government affairs agencies. Without input from the utility operator, it is difficult to assess the extent of impairment and service disruption to any telemetry or to any enterprise control or support circuitry, which may have been impacted.

Restoration and Recovery Estimates

A majority of the damage to OSP resulted from collapsing buildings and potentially engaged multiple contacts with energized electrical conductors as the damage occurred. Contact with commercial power conductors and copper telephony cables usually compromises fine-gauged twisted copper facilities, fusing the facility core into a single unit with no apparent damage to the outer sheath. As such, it is difficult to surmise the serviceability of functional network elements. Cable acceptance testing of those affected elements is advised.

It seems that a high percentage of the aerial OSP was older, possibly built between the late 1970s and the mid-1980s, based on the style and version of pole-mounted facilities, such as terminals, splice cases, cross-connects, control points, and b-box apparatus. It was evident that these OSP facilities suffered from deferred system maintenance, most likely because of budgetary decisions.

We saw several open splice cases, exposing the entire wire contents to wind, rain, heat, and humidity and expediting facility element deterioration of the plastic-insulated conductors (PIC) in telephony cables. In many cases, attachment to support structure (poles) using a mechanical clamping device was deferred by lashing aerial cables and their messenger with hemp or nylon ropes to the concrete poles (see Fig. 5-1).

These maintenance issues are most likely concerned with the age (mortality dates) and condition of the remaining infrastructure and the viability those network elements represent as the utility contemplates continuing wireline network operation to meet service demand.

Our final visit in the telecommunications sector concluded at the site of the former ATS (see Fig, 5-2). With the rubble pile removed exposing the ground floor, the team had an opportunity to peer into the cable vaults (Fig. 5-3 and 5-4). Conceivably, the facilities inside the vault enclosure could provide a starting point for analysis.

Figure 5-1. Telco facilities attached to a pole with rope

Figure 5-2. Collapsed Teleco building in Port-au-Prince (Courtesy of the U.S. Federal Communications Commission)

Figure 5-3. Teleco building site after the rubble was cleared away; back-up generation facilities are visible in the background to the left

Figure 5-4. Teleco building cable vault

5.3 Wireless Telecommunications

As previously noted, the wireless cellular provider Digicel installed its antenna arrays on hillsides, building roofs, and intercessory locations as needed. The system resilience was compromised because of the number of antennas damaged by collapsing buildings. According to the FCC Communications Assessment Team's report, during initial discussions with Digicel, it was estimated that 20 percent of the company's network was damaged beyond repair and unable to return to service. By January 27, 2010, the company restored 92 percent of its RF capability with the regulators grant of additional spectrum for 12 months (DigiCel 2010). Those sites unable to return to service are to be dismantled with reusable parts cannibalized for use at other locations.

Structural damage to the company's control center (MCS) near the Pacolt district was hampered with the damage to the environmental HVAC system supporting the network.

The workarounds performed included re-routing the building's cooling system, turning down non-essential equipment to save electricity, and lowing equipment room ambient temperatures before full restoration could be accomplished.

As post-disaster international calls increased through the Digicel network, the FCC assessment team reported that call blockage occurred, due in part to system network capacity. One resolution was to augment its transport microwave to the Dominican Republic. An additional constraint comes from limited capacity accorded Digicel in the Americas-Region Caribbean Ring System (ARCOS) fiber optic cable landing site in the Dominican Republic.

This chapter relies on the FCC analysis outlining the dependency on the microwave transport across the border into the Dominican Republic to reach the ARCOS fiber. The rate costs quoted wireless providers to access the Port-au-Prince cable landing far exceeded the cost of the construct and operation of the microwave system into the neighboring country. As a result, the opportunity to create a redundant path for international traffic for the wireless carrier is not economically feasible.

5.4 Major Observations and Recommendations

Observation: Wireline Telecommunications

The age and condition of the remaining infrastructure is an issue. The system functionality is constrained by the apparent deferred maintenance and substandard construction techniques, which leads to minimal confidence in the system's ability to meet current service demand and future growth potential. We are confident in the recovery of the BDSNicable landing, once the facility is exposed and tested. Following acceptance testing and reestablishment of the terminating equipment (whether at the current site or other location), Teleco has the opportunity to evaluate the next genesis of its operation and decide whether to continue a blend of fiber and copper network culminating at a tandem and end offices and then delivered to individual consumers. Alternatively, it may extend fiber-optic facilities directly to consumers of advanced services. A twisted pair copper telephony solution either digitally transported and then analog converted to POTS services is not systemically viable or economically feasible in the current OSP condition and the community culture. We would recommend an alternative.

Observation: Wireless Telecommunication

Digicel provides wireless cellular service based on pre-purchased minutes or metered usage, which curbs the opportunity for theft of services. The system is supported on its own transport medium and does not rely on the incumbent telecom utility. A similar dynamic is occurring widely in the United States with customers moving away from reliance on wireline-based systems. If the primary route out of the country is disabled or impaired, operators suffers lost revenues and unmet customer expectations. Operators should give serious consideration to establishing diverse routing of international traffic, thus providing system redundancy and service reliability.

Recommendations

Each of the telecommunication wireline and wireless disciplines discussed have their benefits and challenges. The community presence of both wireline and wireless service offerings provide consumers flexibility, mobility, and choice. Both operators are stand-alone operations and are not reliant of the other. The telecommunications industry experiences business challenges from a variety of sources—the regulatory environment including service obligations and consumer-base expectations, coupled with pressures from political subdivisions and economic issues within the community. As a result, operators must determine the appropriate processes to formulate and develop procedures and construction and maintenance standards in line with operational methodologies to meet those challenges and regulatory requirements. Considering market competition from multiple wireless operators, the effects of the earthquake, and the potentially long-term mobility of individuals forced from their homes and businesses and the devastation of Teleco's wireline network, we suggest the following actions:

- Teleco should be relieved of any obligation to serve all potential customers throughout its current allocated area.[3] It should continue to serve its existing customer base and should not seek to discontinue service in any region or district without probable cause and regulatory approval.
- To ensure consumer pricing stability in the region, Teleco and all current wireless operators should cap and freeze all rate schedules at their current levels until the government and industry stakeholders can determine the future needs and technological path that the telecommunications sector should embrace.
- Due to the current telecommunication service disruptions, network instability, impaired reliability, unreliable recurring revenue, and impaired service assurance, we suggest that current and future discussions center on the convergence of the two telecommunications disciplines—wireline and wireless. A nascent provider could potentially utilize the remaining strengths, resource, and intellectual and capital assets in the wake of the events of January 12. Following any acceptance of the above consolidation, we envision an optically-based landline approach to serve a wireless-based feeder and distribution system into neighborhoods. To ensure recurring revenue, the emerging utility should consider basic service delivery based on pre-pay minutes of use. Advanced services (OC-3 and above) should be delivered under contract with stipulations that require prearranged deposits on user accounts for continuing service delivery.
- Conversely, the emergence of a combined operations partnership, coupled with the pathways of both ARCOS and BDSNi trans-Caribbean networks, could provide redundancy and diverse traffic routing, thus improving customer service and disaster response capabilities.

---

[3] If an allocated area, such as a telephone exchange or wire center, still exists today.

- Further benefits could be realized by employing the intellectual properties and human assets of the wireline and wireless communities at the base level and developing management and occupational skill sets focusing on construction and maintenance standards, service reliability, and a corporate code of conduct.

The final recommendation represents a suggested course of action based on our field observations and review team industry experience,[4] the analysis provided in FCC Assessment Team documents, and industry findings distributed on the Internet.

5.5 References

Digicel Web site <www.digicelgroup.com/en/media-center/press-releases/digicel-in-the-community/digicel-haiti-now-has-more-than-92-of-cell-sites-back-on-air>.

IEE Communications Society Web site <http://dl.comsoc.org/surveys/public/4q99issue/Rahnema.pdf>.

Webopedia Web site <http://www.webopedia.com/TERM/C/CDMA.html>.

Columbus Networks Web site <www.columbus-networks.com >.

[4] Including, but not limited to, telecom OAM and regulatory expertise.

6. Roads and Bridges

6.1 Description of System

Haiti is a relatively small country with a network of roads and highways connecting cities and villages. Most of the roads are asphalt and graded dirt, but some roads in the cities are concrete. Roads conditions before the earthquake were fair at best with many roads heavily potholed and saturated with water.

6.2 Overview of System Performance

In many cases it was difficult to tell earthquake damage from normal road wear and lack of maintenance. However, damage was evident along the coastal areas where road fills had been placed on liquefiable soils, which caused settlement and lateral spreading. Landslides also caused damage by either blocking or undermining roads.

6.3 Site 1 – Highway 3 Near Peligre Dam (N72° 2' 23.6''; W18° 54' 05.6")

There were a number of landslides blocking portions of Highway 3 near the Peligre Dam and reservoir. Many of the smaller landslides blocked only the shoulders. Figures 6-1 and 6-2 show a major landslide being cleared by heavy equipment.

Figure 6-1. Landslide being cleared (Photo by Curt Edwards)

6.4. Site 2 – Highway 1 Near Arcahaie

Highway 1 is a heavily traveled two-lane road along the western coast of Haiti. Figure 6-3 shows the typical cracking in the road we saw frequently as we got closer to the coastline. It is unknown whether the cracks were there before the earthquake or were the result of lateral spreading of the ground towards the sea.

Figure 6-2. Heavy equipment moving landslide blocking road (Photo by Curt Edwards)

Figure 6-3. Road and shoulder cracks (Photo by Curt Edwards)

6.5. Site 3 – Bridge on Highway 2 West of Leogane (N72° 43' 38.23"; W18° 25' 21.9")

This bridge consists of thee simply supported spans with a total length of about 40 m (120 ft.). The bridge is a concrete "T" beam type structure supported on two cast in place piers and two concrete abutments. The beams rest on rubber pads and have no shear keys (Fig. 6-4).

On inspection of the bridge, we found minimal damage caused by mostly longitudinal movement resulting in pounding at the abutments, compression and expansion of the expansion joints, and minor spalling. It also appears that one of the piers was leaning slightly. Traffic was not significantly affected, although the wide gap in the joint caused quite a jolt when driven over. Figures 6.5 through 6.10 show typical damage.

It was surprising that there was so little damage to this bridge because it was so close to the epicenter of the earthquake. Although the bridge will probably not be repaired, it would be easy to jack up the spans and reset the gaps and bearing pads.

Figure 6-4. Highway 1 bridge (Photo by Curt Edwards)

Figure 6-5. Compressed joint with pad out; note the leaning pier (Photo by Curt Edwards)

Figure 6-6. Wide joint with pad out (Photo by Curt Edwards)

Figure 6-7. Damage at abutment (Photo by Curt Edwards)

Figure 6-8. Top view of expansion joint (Photo by Curt Edwards)

Figure 6-9. Spalling at bottom of "T" beam (Photo by Curt Edwards)

Figure 6-10. Typical abutment movement (Photo by Curt Edwards)

6.6. Site 4 – Landslide on Highway 2 (N72° 48’ 50.15”; W 18° 25’ 46.7”)

There were extensive landslides and rockfalls on the cut slopes adjacent to Highway 2. Many of the major blockages had been cleared, but there was still quite a bit of debris blocking the roadway at the time of the investigation. No heavy equipment was seen in the area. There was also a displacement in the road, which possibly indicated failure of the underlying soils (Fig. 6-11 and 6-12).

Figure 6-11. Road cracking at landslide location (Photo by Curt Edwards)

Figure 6-12. Large rocks blocking road (Photo by Curt Edwards)

6.7. Sites 5 and 6 – Highway 2 Damage West of Leogane (N72° 37' 56.9"; W 18° 28' 24.2" and N72° 43' 28.5"; W 18° 25' 23.5")

Sites 5 and 6 along Highway 2 are typical of highway construction on unsuitable underlying soils. About 2 to 3 m (6 to 9 ft.) of highway fill was apparently placed on top of saturated alluvium at two creek crossings. The fill appeared to be well compacted and the road sections seemed adequate. However, the underlying soils liquefied and moved causing lateral spreading and cracking of the fill and pavement sections. This damage mostly occurred at the highway shoulders allowing traffic to continue. Some sections, however, were limited to one lane of travel.

Site 5

Figure 6-13. Severe failure of highway fill (Photo by Curt Edwards)

Figure 6-14. Failure near concrete culvert (Photo by Curt Edwards)

Figure 6-15. Shoulder spreading (Photo by Curt Edwards)

Figure 6-16. Shoulder and road failure (Photo by Curt Edwards)

Figure 6-17. Lateral spread adjacent to highway fill (Photo by Curt Edwards)

Figure 6-18. Dangerous fissure; note road section and fill (Photo by Curt Edwards)

Figure 6-19. Complete highway damage (Photo by Curt Edwards)

Figure 6-20. Cracked wing wall at culvert (Photo by Curt Edwards)

Site 6

Figure 6-21. Vertical displacement of 15 cm (6 in.) (Photo by Curt Edwards)

Figure 6-22. Box culvert with wingwall crack (Photo by Curt Edwards)

Figure 6-23. Shoulder cracking (Photo by Curt Edwards)

Figure 6-24. Lateral spreading adjacent to the highway (Photo by Curt Edwards)

Figure 6-25. Unsafe conditions on both shoulders (Photo by Curt Edwards)

6.8. Site 7 – Landslide on Highway 2 (N72° 48' 30.6"; W 18° 25' 33.4")

This site is located on Highway 2 just west of Grand Goave. It appears that the highway was constructed on an old landslide, which slid further during the earthquake. Figures 6.26 and 6.27 show an overview of conditions on both sides of the damaged road.

The road had a vertical displacement of near 0.3 m (1 ft.) in places (Fig. 6-28); however, it was still passable as a one-lane road.

It was very difficult to tell if there was settlement or movement below the wall. The retaining wall did not appear to be damaged and seemed to move as a unit with the fill (Fig. 6-29 and 6.31). The uphill side of the road showed evidence of spreading indicating slippage of the slope (Fig. 6-30).

Our team continued driving west on Highway 2 to Miragoane. As we headed west, there was less evidence of any earthquake damage, so we stopped. Just outside of Miragoane, we observed heavy equipment constructing a new portion of Highway 2, which bypasses a segment that was flooded by a lake overflowing during major hurricanes a few years ago. The temporary dirt road was passable, and the heavy equipment assigned to this construction would have been better suited to the Port-au-Prince debris removal.

Figure 6-26. Road fill movement; note rocks on the left (Photo by Curt Edwards)

Figure 6-27. Overview (Photo by Curt Edwards)

Figure 6-28. Road displacement and warning sign (Photo by Curt Edwards)

Figure 6-29. Undamaged retaining wall (Photo by Curt Edwards)

Figure 6-30. Uphill slope showing spreading (Photo by Curt Edwards)

Figure 6-31. Slope below retaining wall (Photo by Curt Edwards)

6.9. Site 8 – Landslide at Road in Boutillier (N72.320043°; W 18.4989°)

There was a large landslide in the Boutillier area in the mountains above Port-au-Prince. It was at least 300 m (300 yards) long and had a scarp of at least 20 m (60 ft.). The slide eroded most of the paved road leading to the top of the mountain and the communication facilities there (Fig. 6-32 and 6.33). A temporary dirt bypass road had been constructed just above the old damaged road (Fig. 6-34). The slide also damaged two houses directly below the road and businesses in the area (Fig. 6-35, 6.36, and 6.37).

Figure 6-32. Slide looking west (Photo by Curt Edwards)

Figure 6-33. Slide looking east (Photo by Curt Edwards)

Figure 6-34. Temporary bypass road (Photo by Curt Edwards)

Figure 6-35. Houses crushed by landslide (Photo by Curt Edwards)

Figure 6-36. Extent of slide (Photo by Curt Edwards)

Figure 6-37. Slide from below (Photo by Curt Edwards)

6.10. Major Observations and Lessons Learned

It was surprising that there was no major bridge damaged even though many were located in high-risk areas, such as coastal river crossings. Although there was some cracking, span movement, abutment settlement, and minimal repairs, the bridges were still functional. This allowed for continued movement of rescue vehicles, aid packages, and critical vehicles such as water delivery trucks. If one or more of these bridges had collapsed, the already heavy traffic would have been worse, further delaying response times.

Some of the lessons learned from this event include:

1. Reassign available heavy equipment from less important tasks to road clearing and debris removal. This will greatly improve emergency response times and improve recovery.
2. Evaluate ground conditions during highway construction. If possible, mitigate hazards such as landslides and liquefaction/lateral spreading zones to avoid road damage.

7. Ports, Marine Oil Terminals, and Airports

Port, marine oil terminal, and airport facilities were investigated during the ASCE-TCLEE post-earthquake reconnaissance following the January 2010 Haiti earthquake. Many of the waterfront facilities near Port-au-Prince suffered significant damage during the earthquake and aftershocks, with the majority of the structures damaged and many unusable for weeks or more following the earthquake due to strong ground shaking and liquefaction.

Given the large humanitarian needs, the airport underwent emergency repairs and was partially operational by January 13. The port of Port-au-Prince was heavily damaged and only partially operational with larger ships offloading to smaller, self-unloading ships and the use of temporary dock barges. Some international humanitarian aid was also routed through smaller private ports near Port-au-Prince.

7.1 Description of Systems

Port Facilities

Several port facilities (Fig. 7-1) are located near Port-au-Prince:

- **Port de Port-au-Prince (PPAP):** Terminals located in Port-au-Prince consisted of a finger pier, a marginal wharf, and two pile-supported roll-on/roll-off (Ro/Ro) berths that handled container and break bulk cargo.
- **Terminal Varreaux:** Located north of Port-au-Prince, this terminal included a marginal wharf with a finger pier extension that handled break bulk materials.
- **Cimenterie National (CINA):** CINA included facilities north of Port-au-Prince at a cement plant to handle break bulk materials for the cement plant consisting of a finger pier and a bulkhead Ro/Ro berth.
- **Moulins d'Haiti:** This included facilities at a flour mill north of Port-au-Prince to handle break bulk materials for the flour mill. It consisted of two bulkheads, one serving as a Ro/Ro berth.
- **Terminal Abraham:** This terminal included a finger pier located west of Port-au-Prince for break bulk cargo.

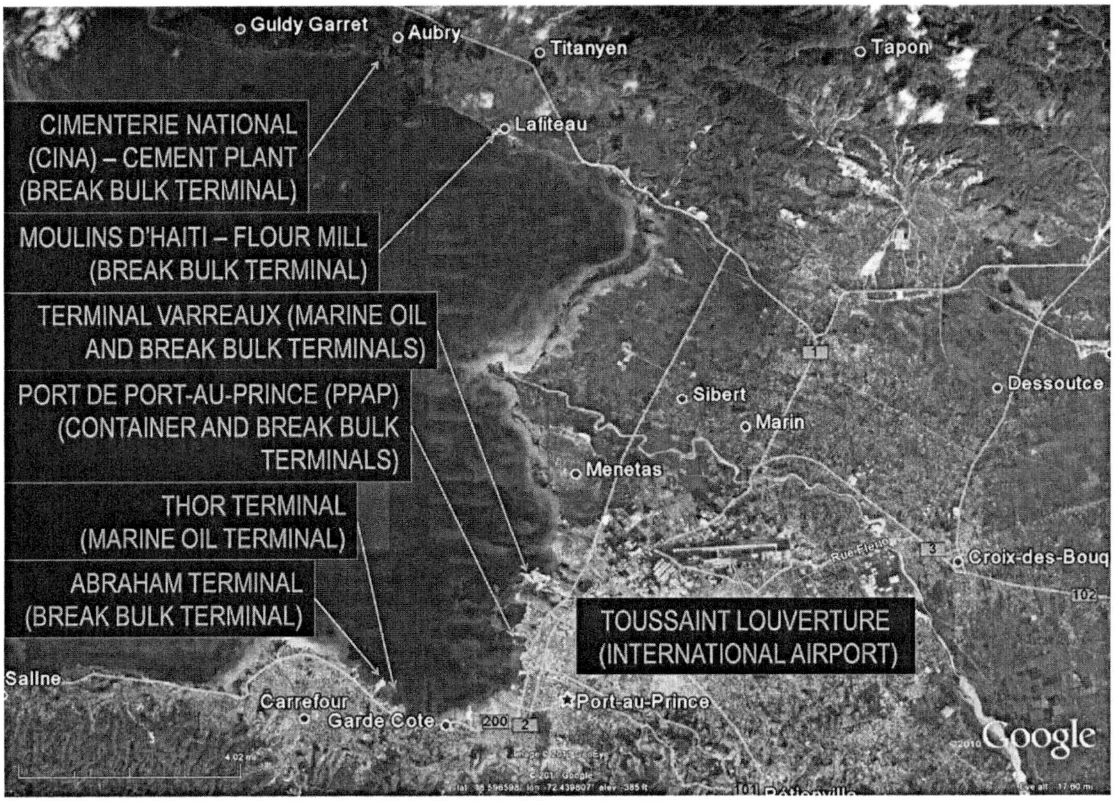

Figure 7-1. Overview of investigated facilities in the Port-au-Prince area (Image © 2011 Google, GeoEye, with permission from GeoEye)

Marine Oil Terminals

Two marine oil terminals near Port-au-Prince that handled the majority of the country's petroleum imports (Fig. 7-1):

- **Terminal Varreaux Marine Oil Terminal:** This terminal was located north of Port-au-Prince and adjacent to the break bulk facilities at the same terminal. The facilities consisted of a finger pier, pipeline system, storage tanks, and a distribution system.
- **Thor Marine Oil Terminal:** Located west of Port-au-Prince, this terminal consisted of an earth mole with a finger pier extension, pipeline system, storage tanks, and a distribution facility.

Airports

There was one major airport near Port-au-Prince:

- **Toussaint Louverture International** (Fig. 7-1) is one of two international airports in Haiti; the other is located in Cap Haitian on the north side of the country. The airport included a single runway about 3 km (10,000 ft.) long.

7.2 Overview of System Performance

The waterfront facilities (ports and marine oil terminals) and the airport facility were significantly damaged during the earthquake. The airport damage was mostly structural due to the strong ground shaking. The waterfront facilities were damaged primarily due to liquefaction of the foundation and backfill soils, although there were also indications that many of the facilities were not designed to current seismic standards or were corroded or damaged before the earthquake and may have suffered significant damage due to inertial loading. There was also damage to backland storage tanks and pipeline systems at the marine oil terminals.

The loss of both the airport and waterfront facilities greatly hampered the humanitarian and post-recovery efforts. The functionality of the airport was partially restored within about 1 day; however, air transportation was not able to accommodate the large volumes of cargo necessary for the large-scale humanitarian and post-earthquake recovery efforts needed after such a large disaster. It was necessary to restore the waterfront facilities to provide this needed capacity.

Port Facilities

Port de Port-au-Prince (PPAP)

Haiti's main seaport is located in the capital of city of Port-au-Prince about 1 mile south of the city of La Saline and 1 mile west of Croix des Bossales. The port is centered at a latitude of 18.5561 degrees and longitude of about -72.3487 degrees, which is about 22 km (13.5 miles) from the earthquake epicenter. There was no direct measurement of ground shaking at the port. Based on the preponderance of liquefaction, however, it is estimated the peak ground acceleration was about 0.2 g or greater.

The port is operated by the government's Authorite Portuaire National (APN). APN is a landlord port, whose current functions are both regulatory and operational. It was the largest and busiest container port in Haiti, handling about 1200 containers per day and 170,000 TEU's[5] per year before the earthquake, according to APN officials. The port primarily consists of two separate facilities designated as the north wharf and the south pier (Fig. 7-2), both built between1978 and 1980. There were seven berths including two Ro/Ro berths before the earthquake.

Port Entrance

The main port entrance was significantly damaged during the earthquake and became unusable afterwards. Liquefaction evidence and lateral spreading were observed along the port entrance road (Fig. 7-3).

---

[5] A TEU is a standard shipping unit of measure, and refers to a standard 6-m (20-ft.) equivalent unit (TEU) shipping container.

Figure 7-2. PPAP Facilities prior to the Earthquake
(Image © 2011 Google, GeoEye, with permission from GeoEye)

North Wharf

The North Wharf was a marginal wharf constructed in 1979 approximately 457 m (1,500 ft.) long and 12 m (70 ft.) wide supported by 45 cm (18 in.) square pre-stressed concrete piles with plumb and two batter piles per bent (Fig. 7-4, 7-5). There were five plumb piles and two batter piles per bent (NAVFAC 1994). The wharf supported a rail-mounted gantry container crane and two mobile harbor cranes. The water depths along the berth line ranged from 10 to 12 m (33 to 39 ft.) before the earthquake. A Ro/Ro ramp was located adjacent to the wharf on the east end.

The north wharf and Ro/Ro ramp collapsed during the earthquake (Fig. 7-4), and two of the three cranes were submerged in the water. During the earthquake, two vessels were moored to the north wharf, and a crew member from one vessel took several photographs of the immediate post-earthquake damage. Figures 7-6 through 7-9 are photographs taken from the moored ship following the earthquake. The photographs provide evidence that the wharf did not initially collapse completely: After strong shaking the wharf was still above water and gradually collapsed into the bay. Figure 7-6 provides an aerial view of the North Wharf several days after the earthquake and shows complete collapse of the wharf into the bay.

Figure 7-3. PPAP port entrance damage
(Photo courtesy of Nason McCullough)

Figure 7-4. PPAP North Wharf and South Pier before and after the earthquake (Image © 2011 Google, GeoEye, with permission from GeoEye)

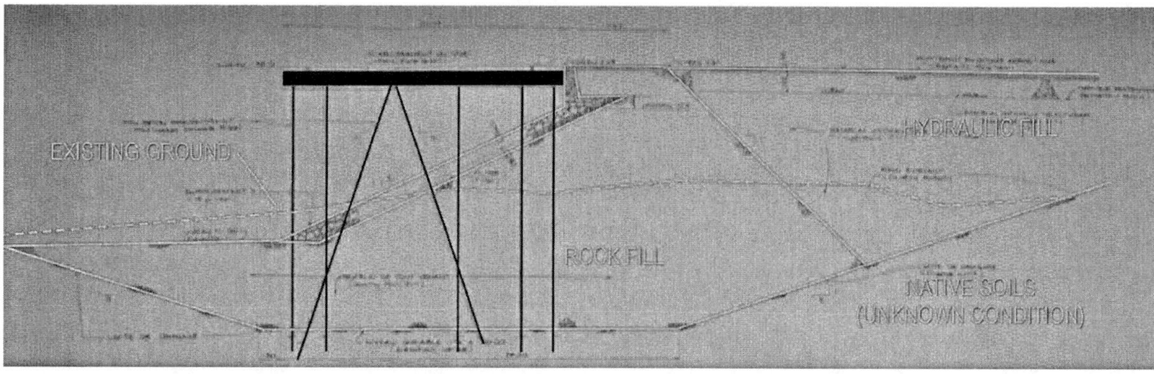

Figure 7-5. PPAP North Wharf as-designed geometry

Figure 7-6. PPAP North Wharf damage 4 days after the earthquake (Photo courtesy of the Department of Defense)

There was evidence of liquefaction in the backland areas (Fig. 7-4). Large ground cracks were also observed along and near the edge of the waterline (Fig. 7-7 through 7-8), presumed to be caused by the significant liquefaction.

Two steel-frame warehouses suffered significant damage, but did not collapse (Fig. 7-9). The steel trusses were severely buckled in several locations, and the foundation supports were also severely damaged or separated (Fig. 7-10) from the columns in several locations. Figure 7-11 shows the significant internal damage to the warehouse.

South Pier

The South Pier was a pile-supported finger pier constructed in 1979 about 380 m (1250 ft.) long and 18 m (60 ft.) wide (Fig. 7-4). The pier was pile supported with two plumb and four interior batter piles per bent (Fig. 7-12). Two pile-supported causeways, a vehicular bridge, and a pedestrian bridge connected the pier to an island where port security was located. The western end of the pier was connected to four dolphins by walkways.

Figure 7-7. PPAP North Wharf damage and ground cracks (looking west) (Photo courtesy of Alex Augustin)

Figure 7-8. PPAP North Wharf damage and ground cracks (looking east) (Photo courtesy of William Bruin)

Figure 7-9. PPAP North Wharf warehouse damage (looking northwest) (Photo courtesy of Stu Werner)

Figure 7-10. PPAP North Wharf warehouse damage (footing destroyed)
(Photo courtesy of Alex Augustin)

Figure 7-11. PPAP North Wharf warehouse damage
(Photo courtesy of William Bruin)

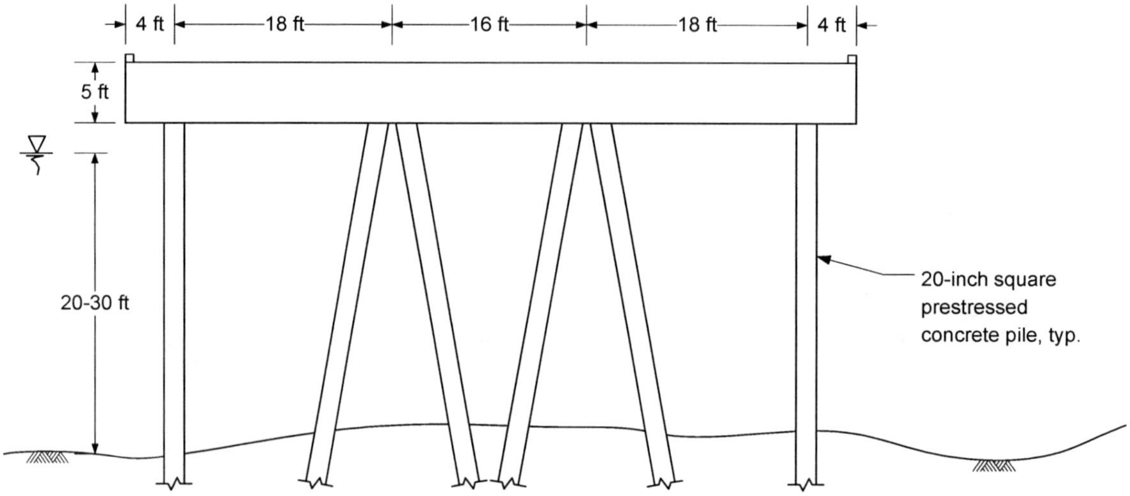

Figure 7-12. PPAP South Pier geometry (not to scale)

There was evidence of the pier being in disrepair prior to the earthquake. Figure 7-13 shows an aerial image as well as several inspection photographs of the South Pier prior to the earthquake. Prior to the earthquake the western most walkway was already missing and there was noted damage to the middle catwalk and the pedestrian causeway from vessel impact. Due to the damage, the pedestrian causeway was not structurally connected to the South Pier.

During the earthquake about 120 m (400 ft.) of the water-side end of the pier collapsed west of the vehicular causeway. Additional portions of the walkways also collapsed. Several ships were moored to the dolphins and the western segment of the pier during the earthquake. There was also noted liquefaction and lateral spreading of soil along the perimeter of the island, causing lateral movement and differential settlement of the causeway abutments (Fig. 7-14). Figure 7-14 also shows damage to the piles, which likely occurred before the earthquake as evidenced by the considerable corrosion of the rebar and deteriorated concrete spalling.

Photographs taken immediately following the earthquake suggest that the South Pier did not collapse during the initial strong ground shaking; however, this has not been verified. Although the pier survived the main shock, the numerous aftershocks in the region could have caused the actual collapse.

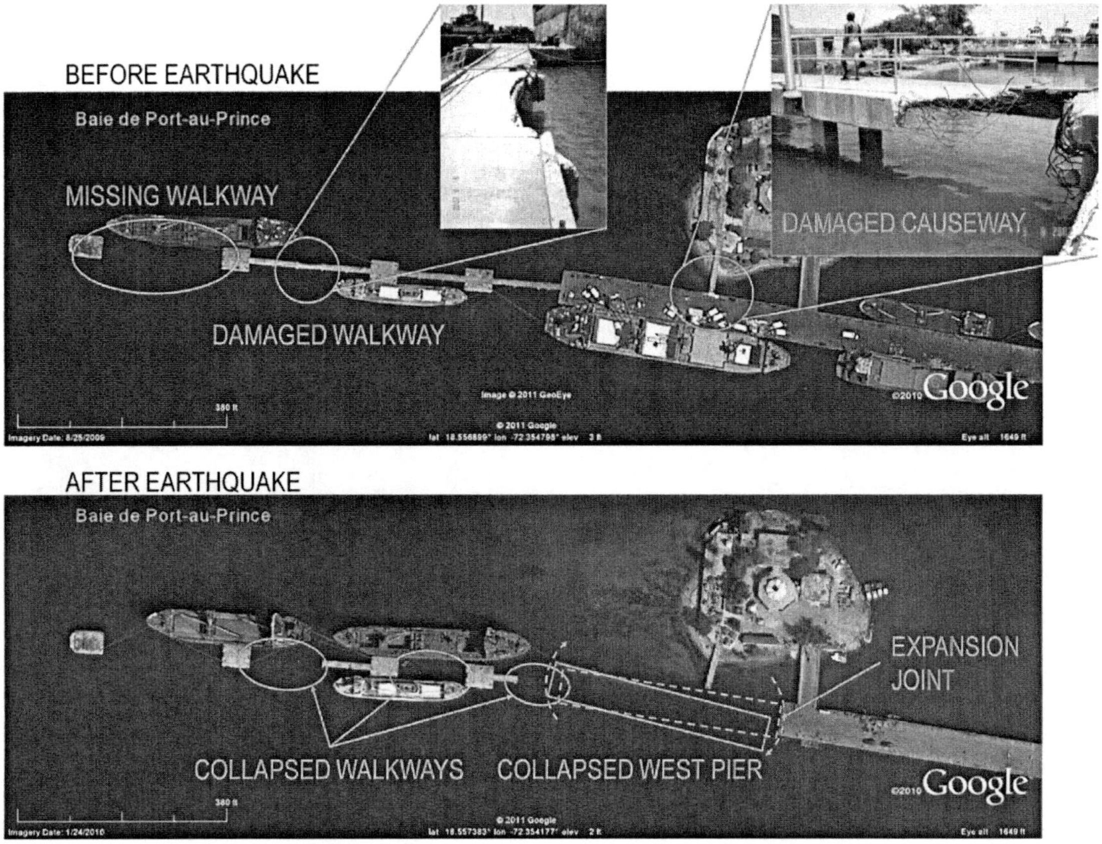

Figure 7-13. PPAP South Pier damage before and after earthquake (Photos courtesy of Dave Mock, reproduced by permission; Images © 2011 Google, GeoEye, with permission from GeoEye)

Figure 7-14. PPAP South Pier damage – pedestrian causeway (Photos courtesy of Nason McCullough and Alex Augustin)

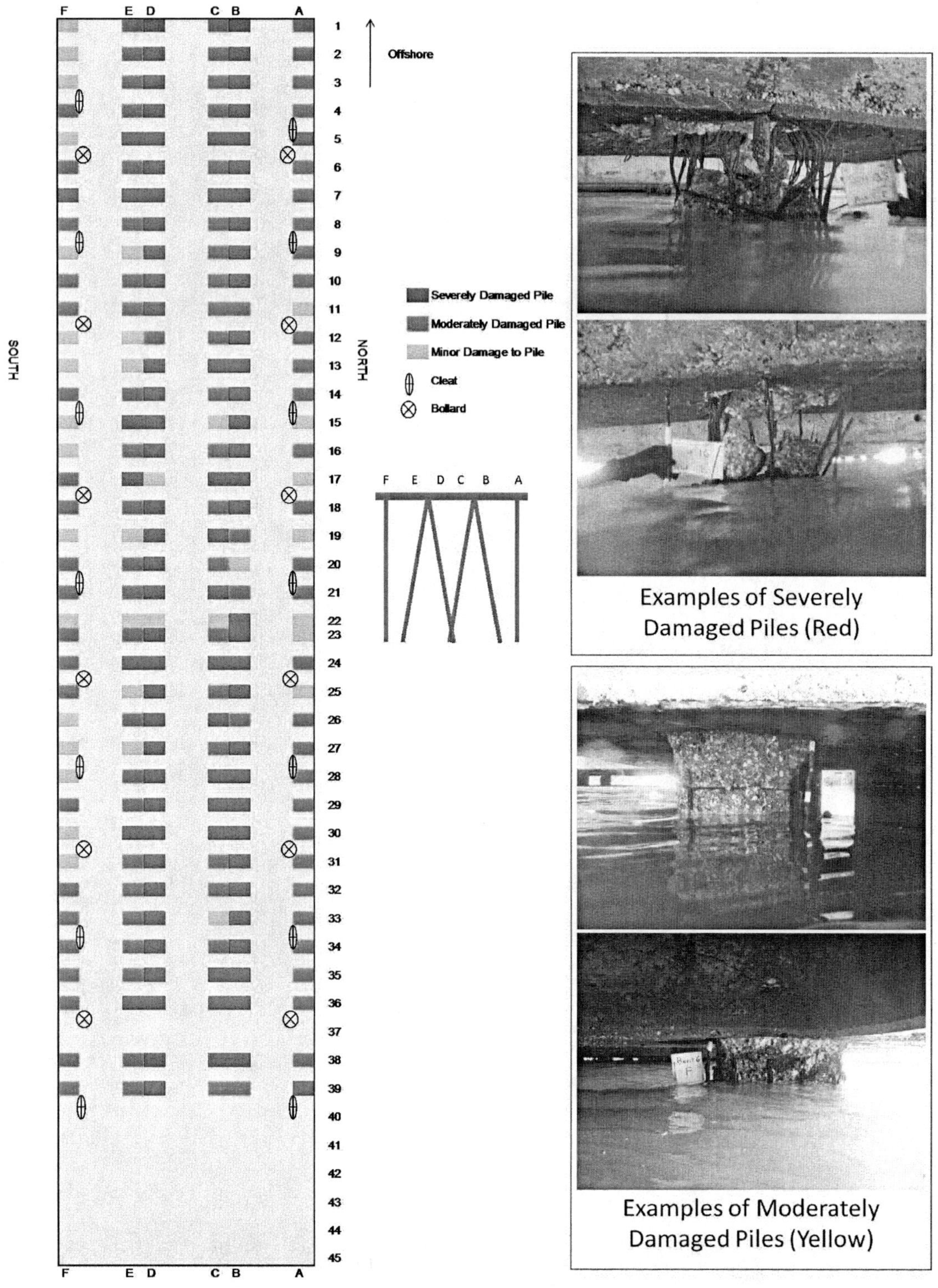

Figure 7-15. PPAP South Pier damage, not to scale (Figure and photo courtesy of the Department of Defense)

The portion of the South Pier that remained standing after the earthquake was inspected by the U.S. military and was observed to have considerable damage due to the earthquake, but also possibly due to damage prior to the earthquake. The U.S. military documented the damage to the piles, and as illustrated in Figure 7-15, 143 of the 156 batter piles were damaged (yellow and red colored piles) and that 90 of the 165 batter piles failed completely (the darker shaded piles). It is also of note that the vast majority of the batter piles that failed completely were on the north side of the pier. This may be due to directionality of the earthquake ground motions, restraint provided by the vehicular causeway, or another unknown reason.

It is uncertain why 260 m (850 ft.) of the South Pier remained standing while the western 120 m (400 ft.) collapse. It might be the lateral support provided by the causeway, which connected the remaining portion of the South Pier to the island. The collapsed portion of the South Pier was not connected to the island (Fig. 7-13). However, there may have been other reasons contributing to the collapse, such as damage before the earthquake or another unknown reason.

The U.S. military also inspected the dolphins and observed a similar level of damage, as shown in Figure 7-16. Note that mooring Dolphin #1 is the farther offshore, and Dolphin #4 is the closest to shore.

Grain Silos

There were six grain silos constructed in 2009. The silos were visibly undamaged, though there was uplift of several inches observed for one of the silo foundations (Fig. 7-17). Surface evidence of liquefaction was observed near the grain silos but not in the immediate vicinity. It is possible that the silos' foundations included soil improvement or that they were supported on deep foundations.

Reclaimed Beaches

Several beaches on the northern end of the port had recently been reclaimed land. Significant lateral spreading and ground cracks were observed in these beaches. Figure 7-18 shows the observed damage to the beaches.

Backland

There were areas of liquefaction in the backland areas of the port. Figure 7-19 shows liquefaction induced settlement of backland pavements adjacent to a culvert.

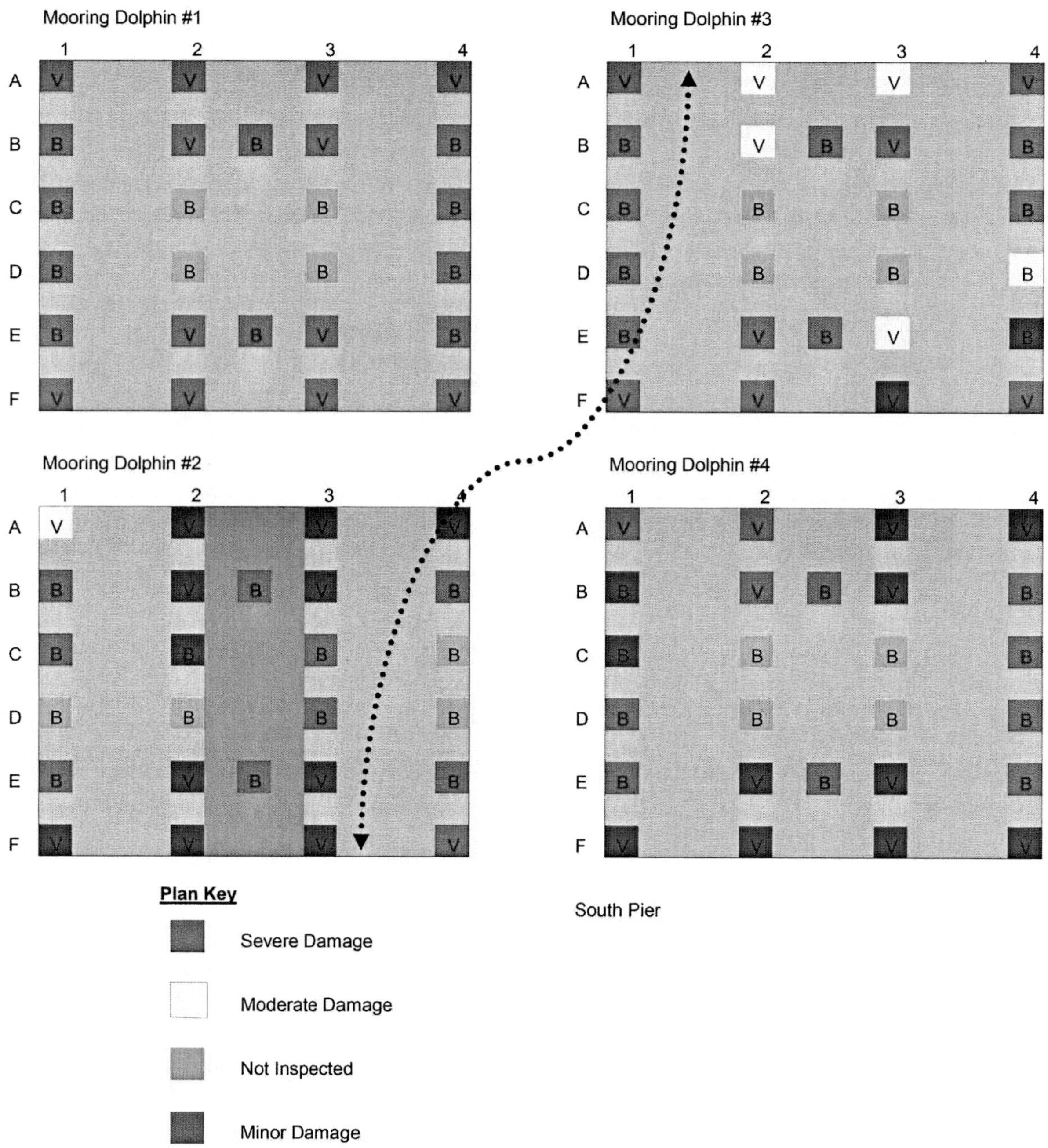

Figure 7-16. PPAP South Pier dolphin damage, not to scale (B=Batter, V=Plumb) (Figure courtesy of the Department of Defense)

Figure 7-17. PPAP grain silo foundation uplift (Photo courtesy of William Bruin)

Figure 7-18. PPAP reclaimed beach damage (note the grain silos in the background) (Photo courtesy of the Department of Defense)

Figure 7-19. PPAP backland damage (Photo courtesy of Nason McCullough)

Varreaux Break Bulk Terminal

Varreaux Terminal is privately owned and operated by WinGROUP and is located just outside Port-au-Prince in the city of Cite Soleil (18.574°N, 72.345°W). The break bulk portion of the terminal consisted of a 60 m (200 ft.) long by 9 m (30 ft.) wide pile-support marginal wharf with a 109 m (360 ft.) long by 24 m (80 ft.) wide pile-supported finger pier (Fig. 7-20), which was constructed in the early 1990s. A second finger pier, used primarily for the marine oil terminal, was approximately70 m (230 ft.) long and 7 m (25 ft.) wide. The wharf/piers had a draft of approximately 9 to 11 m (30 to 36 ft.). The terminal was located about 23 km (14.5 miles) from the earthquake epicenter. It is estimated that the site was subjected to a shaking intensity of MMI VII or VIII based on U.S. Geological Survey (USGS) map data.

Figure 7-20. Terminal Varreaux (top: before earthquake, bottom: after earthquake)
(Images © 2011 Google, GeoEye, with permission from GeoEye)

The break-bulk terminal was supported on 60-cm (24-in.) diameter steel pipe piles, with concrete caps and a concrete deck. The entire finger pier portion of the break-bulk terminal collapsed during the earthquake, killing 56 people when the wharf collapsed while a ship was being unloaded. Ground cracks and lateral spreading were noted behind the marginal wharf (Fig. 7-21), which resulted in movement and damage. The piles of the remaining marginal wharf were observed to have separated from the pile caps, and the pile caps had separated from the wharf deck (Fig. 7-22).

Cimenterie Nationale (CINA) Cement Plant

The Cimenterie Nationale (CINA) cement plant (formerly Le Ciment d'Haiti) was located near Cabaret, about 19 km (12 miles) northwest of Port-au-Prince and about 32 km (20 miles) from the earthquake epicenter. The CINA marine terminal (18.711°N, 72.389°W), also known as Quai de Fond-Mombin, is privately operated and consists of a 145-m (480-ft.) long main pier and an adjacent roll-on/roll-off (Ro/Ro) ramp. The terminal handles the importation of bulk cement and cement admixtures for the nearby cement plant. It is estimated that the site was subjected to ground shaking of VII or VIII on the Modified Mercalli Intensity scale.

The CINA main pier (Fig. 7-23) was reportedly constructed in the 1970s and had an average water depth alongside of 8 m (26 ft.). The pier deck was made of 1-m (3-ft.) thick reinforced concrete. The deck supported two large hoppers on rails and a portable conveyor feeder. The shallow portion of the pier was supported on steel sheetpile cellular gravity structures about 12 m (40 ft.) in diameter. The deep-water portion of the pier was supported by two-pile bents with battered 36-inch steel pipe piles.

Adjacent to the pier was an approximately 14 m (45 ft.) wide by 9 m (28 ft.) long Ro/Ro ramp, which was reported to have been constructed in the 1980s. The ramp deck was about a 20-cm (8-in.) thick marginal wharf supported on 91-cm (36-in.) diameter concrete piles with five bents of three piles per bent. Riprap up to a 122-cm (48-in.) diameter was positioned as erosion protection on the slope beneath the ramp.

In general, the waterfront structures performed well, despite evidence of liquefaction and about 20cm (8 in.) of settlement of the onshore approaches to both the pier and the ramp (Fig. 7-24). One of the sheet pile cofferdams was observed to have partially failed (Fig. 7-25); the damage was attributed to the earthquake but not confirmed. Pictures from a previous reconnaissance conducted by Eduardo Fierro of BFP Engineers, Inc. on January 19, 2010, show the cofferdam partially opened with a portion of the sheet piling extending into the berthing area. At the time of the TCLEE visit, the damaged sheet piling had been removed. It was unclear whether there was any loss of material from the cofferdam. However, a slight sag in the pier deck could be observed above the damaged cofferdam as well as a significant transverse cracking of the deck where the sagging began.

Figure 7-21. Terminal Varreaux ground cracks and lateral displacement (Photos courtesy of the Department of Defense)

Figure 7-22. Terminal Varreaux – marginal wharf damage (Photos courtesy of William Bruin and Stu Werner)

Figure 7-23. Cimenterie Nationale (CINA) (Image © 2011 Google, GeoEye, with permission from GeoEye)

Figure 7-24. Settlement at CINA ro/ro ramp
(Photo courtesy of Eduardo Fierro, reproduced by permission)

Figure 7-25. Failure of the cofferdam sheet piling at the CINA Main Pier
(Photo courtesy of Eduardo Fierro, reproduced by permission)

Figure 7-26. Twisting of elevator conveyor segments
(Photo courtesy of Eduardo Fierro, reproduced by permission)

Only minor instances of shaking damage were noted with the main plant area. The most significant was the twisting and collapse of segments of an elevated conveyor that ran from the pier to the cement plant (Fig. 7-26). The conveyor, which was no longer used, was observed to have twisted off its supports and some segments had collapsed completely. At the time of the reconnaissance, many of the dislocated segments had been removed.

Other damage observed included the following:

- There was slope instability at a steep cut adjacent to the plant.
- Concrete frame structures situated on the top of tall concrete silos collapsed (Fig. 7-27).
- The top of a steel kiln stack partially collapsed (Fig. 7-28).
- There was evidence of the uplift of a partially anchored 15,000 gallon diesel fuel tank (about 3 m [10 ft.] diameter, 7 m [24 ft.] tall) (Fig. 7-29). Of the eight anchor bolt locations, four anchors were not installed, and the four remaining had pulled out. However, the tank was not noted to have leaked and was standing after the earthquake.
- Minor cracking damage of masonry infill of concrete building structures throughout the plant was also observed.

Figure 7-27. Collapse of concrete frame structure on top of tall concrete silos (Photo courtesy of Eduardo Fierro, reproduced by permission)

Figure 7-28. Steel stack with partial collapse of top (Photo courtesy of Eduardo Fierro, reproduced by permission)

Figure 7-29. Tank uplift of partially anchored 15,000 gallon diesel fuel tank (Photos courtesy of Eduardo Fierro, reproduced by permission)

Les Moulins d'Haiti Flour Mill

Les Moulins d'Haiti Flour Mill was located near Lafiteau, about 16 km (10 miles) northwest of Port-au-Prince and about 32 km (20 miles) from the earthquake epicenter. The mill primarily consisted of a waterfront terminal, a warehouse, storage silos, a flour mill, and an administration building (Fig. 7-30). The break-bulk terminal (18.691°N, 72.357°W) is privately owned and operated and consisted of a 100 ft. long sheetpile bulkhead constructed in the 1970s (west bulkhead) and an approximately 260 km (850 ft.) long sheetpile bulkhead constructed in 2009 (east bulkhead) (Fig. 7-31). The water depth at the face of the bulkhead was about 10 m (33 ft.). The terminal handled imports for the flour mill. It is estimated that the site was subjected to ground shaking of VII or VIII on the Modified Mercalli Intensity scale.

The west bulkhead consisted of coal tar epoxy coated steel sheetpile sections (Z-sections about 53 cm [21 in.] wide by about 30 cm [12 in.] deep) with a 1 m (3.5 ft.) wide concrete cap. It appears that the west bulkhead was partially backfilled with concrete fill, and significant corrosion was noted of the sheetpile sections. A ground crack approximately 3.8 cm (1.5 in.) wide, measured at the back of the concrete cap, indicated lateral movement during the earthquake. In addition, some bulging of the sheetpile walls was noted in an area of significant corrosion where the concrete cap did not encapsulate portions of the sheetpile above the sea level (Fig. 7-31). However, the west bulkhead was operational after the earthquake.

The east bulkhead consisted of steel sheetpile section (U-sections about (31 in.) wide by about 33 cm [13 in.] deep). There was a large crack (about 20 to 25 cm [8 to 10 in.]) observed in the concrete cap (Fig. 7-32), as well as ground cracks and settlement in the backland behind the bulkhead. Figure 7-33 shows observed outward rotation of the east bulkhead abutment. Additional fill was placed after the earthquake to bring the bulkhead back into operation.

Additional earthquake performance and damage noted at the flour mill included the following:

- Settlement and lateral movement of the warehouse foundations. Differential lateral movement between the footings and floor slab was measured between 1 and 7 cm (3 in.), and differential settlement between the footings and floor slab was measured between 5 to 7 cm (2 to 3 in.).
- Engaging of the anchors for the silos. Of the 20 silos, the two adjacent to the flour mill collapsed with the mill. Silos that were mostly full or full had noted cracking and spalling of the concrete cover over the seismic anchors, with some of the anchors broken (Fig. 7-34).
- The six-story flour mill completely collapsed, killing 13 workers and causing the two adjacent silos to also collapse.
- There was damage to the infilled walls and some of the columns in the administration building.

Figure 7-30. Les Moulins d'Haiti Flour Mill (Image © 2011 Google)

Figure 7-31. Les Moulins d'Haiti Flour Mill – west bulkhead (Photo courtesy of William Bruin)

Figure 7-32. Les Moulins d'Haiti Flour Mill – east bulkhead (Photo courtesy of Nason McCullough)

Figure 7-33. Les Moulins d'Haiti Flour Mill – east bulkhead (Photo courtesy of Nason McCullough)

Figure 7-34. Les Moulins d'Haiti Flour Mill – silo anchors
(Photos courtesy of Nason McCullough)

Terminal Abraham

Terminal Abraham was located near Carrefour, about 4 km (2.5 miles) west of Port-au-Prince and about 17 km (11 miles) from the earthquake epicenter. The break-bulk terminal (18.5451°N, 72.388°W) is privately owned and operated and consisted of an approximately 115 m (380 ft.) long by 9 m (30 ft.) wide pile supported finger pier (Fig. 7-35). It is estimated that the site was subjected to ground shaking of VIII or IX on the Modified Mercalli Intensity scale.

The TCLEE team did not inspect this pier. The following information is courtesy of the U.S. military. The finger pile was supported on 60-cm (24-in.) diameter steel pipe piles filled with concrete and had two piles per bent for 21 bents. The deck was about 30 cm (12 in.) thick. Two batter piles were located at the landside portion to provide lateral resistance. During the earthquake ground cracks and lateral spreading were indicated and resulted in about a 6-m (20-ft.) separation between the shore and the pier and collapse of the first bent. Batter piles were also noted to have punched through the deck.

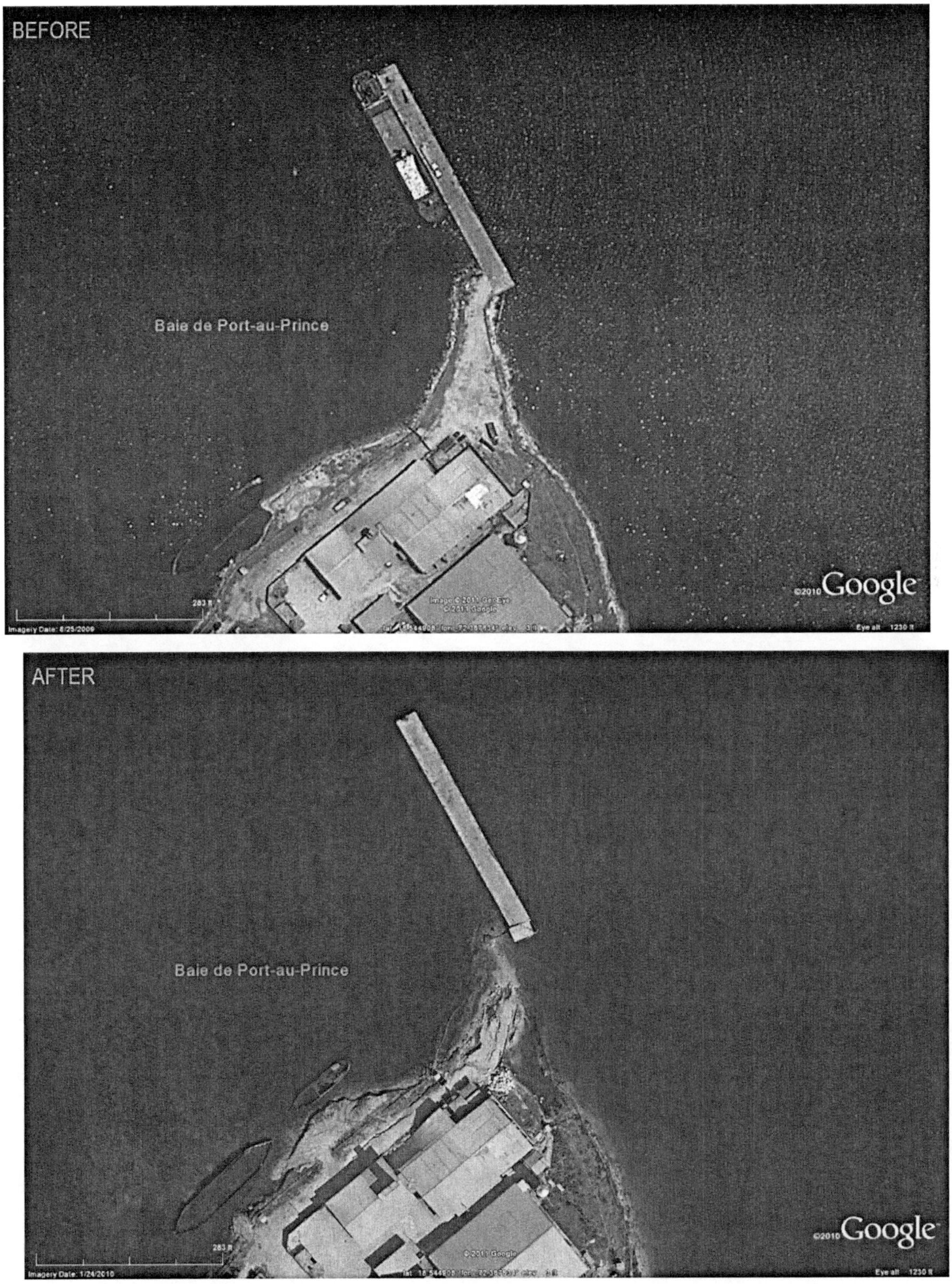

Figure 7-35. Terminal Abraham
(Image © 2011 Google, GeoEye, with permission from GeoEye)

Marine Oil Terminals

Terminal Varreaux Marine Oil Terminal

The Terminal Varreaux Marine Oil Terminal was located adjacent to the Terminal Varreaux break-bulk terminal (previously described) (Fig. 7-20). In addition, the terminal consists of petroleum storage tanks and a distribution system.

The earthquake performance and damage of the terminal included the following:

- The finger pier was heavily corroded before the earthquake (Fig. 7-36), and though there was some additional damage during the earthquake, the pier did not collapse. Figure 7-37 shows some pounding and differential vertical movement of the wharf deck and the punching of a batter pile through the deck.
- There was about 30 cm (12 in.) of settlement at the transition between the terminal and backland, resulting in damage to several of the pipelines.
- There was tank and elephant buckling noted on several of the petroleum storage tanks (Fig. 7-38).

Figure 7-36. Terminal Varreaux Marine Oil Terminal – Marine Oil Terminal finger pier (Photo courtesy of the Department of Defense)

Figure 7-37. Terminal Varreaux Marine Oil Terminal – finger pier damage (Photos courtesy of the Department of Defense)

Figure 7-38. Terminal Varreaux Marine Oil Terminal – petroleum storage tank damage (Photos courtesy of Nason McCullough)

Thor Marine Oil Terminal

The Thor Marine Oil Terminal was located near Carrefour, about 4 km (2.5 miles) west of Port-au-Prince and about 17 km (11 miles) from the earthquake epicenter (18.540°N, 72.387°W). The marine oil terminal is privately owned and operated and included an earthen mole and finger pier, on-land petroleum storage tanks, and a distribution system (Fig. 7-39). It is estimated that the site was subjected to ground shaking of VIII or IX on the Modified Mercalli Intensity scale.

The waterfront structures included an earthen mole about 243 m (800 ft.) long (constructed in the 1920s) that was connected to a pile-supported finger pier approximately 120 m (400 ft.) long and 3.6 m (12 ft.) wide (constructed in the 1950s). During the earthquake, the earthen mole experienced significant lateral spreading attributed to liquefaction, resulting in approximately1.4 m (4.5 ft.) of settlement and the mole increasing in width from 6 m (20 ft.) to around 9 m (30 ft.) wide (Fig. 7-40). The finger pier was damaged before the earthquake. Several piles were broken, and the last bent of the pier collapsed due to ship impact in 2008. The earthquake damage to the pier included differential transverse lateral movement of some precast deck panels of up to 60 cm (2 ft.) (Fig. 7-41).

Figure 7-39. Thor Marine Oil Terminal (Image © 2011 Google)

Figure 7-40. Thor Marine Oil Terminal – damaged earthen mole (Photo courtesy of Nason McCullough)

Figure 7-41. Thor Marine Oil Terminal – finger pier
(Photos courtesy of Nason McCullough)

Two of the petroleum storage tanks were damaged during the earthquake:

- Tank #5 was noted to have elephant foot buckling and cracking of the concrete foundation, though the tank did not release any contents.
- Tank #6 was noted to have tank shell and elephant foot buckling. A weld seam at the base of the tank failed during the earthquake, and the entire contents of the tank were released. The failure of the welds occurred at locations of overlapping plate welds (Fig. 7-42).

Airports

Toussaint l'Ouverture International Airport

The Toussaint l'Ouverture International Airport was located on the north side of Port-au-Prince (18.580°N, 72.291°W), about 27 km (17 miles) from the epicenter. It is estimated that the site was subjected to ground shaking of VII or VIII on the Modified Mercalli Intensity scale.

The air control tower and terminal were heavily damaged during the earthquake and were condemned.

Figure 7-42. Thor Marine Oil Terminal – Tank #6
(Photos courtesy of Nason McCullough)

7.3 Post-Earthquake Repair and Recovery

Port de Port-Au-Prince

Because of the major casualties and destruction of the built infrastructure in the areas affected by the earthquake, it was essential for humanitarian cargoes (medical, food, water, etc.) and commercial cargoes to be transported into Haiti as soon as possible. As the largest seaport in Haiti, PPAP was essential to these emergency relief efforts. However, because of severe earthquake damage, PPAP was unable to receive these cargoes.

To address this situation, a joint task force (JTF) comprised of contingents from the U.S. Navy (Underwater Construction Team One [UCT-1], Naval Mobile Construction Battalion Seven [NMCB-7/Seabees], Naval Beach Group Two, Amphibious Construction Battalion Two [ACB-2], and NAVFAC) and the U.S. Army (7th Sustainment Brigade, 688th Rapid Port Opening Element, and 544th Engineer Dive Team) was deployed to PPAP with initial personnel arriving six days after the earthquake. The joint task force mission was to coordinate shipping into PPAP and to restore shipping operations for support of humanitarian emergency response and recovery efforts. Task force activities were closely coordinated with APN.

On January 19, 2010, HNLMS Pelikaan *(A804)*, a Royal Netherlands Navy logistic support vessel, was the first ship to enter the harbor following the earthquake. After a hydrographical survey of the harbor, a route was cleared using the ship's crane, and HNLMS Pelikaan could berth and unload its cargo and marines. On January 21, it was announced by the Dutch Ministry of Defense that HNLMS Pelikaan had finished unloading its cargo and would depart for Guantanamo Bay Naval Base. The French vessel Francis Garnier was the first vessel to berth at APN's damaged South Pier on January 21, and it was able to assist in the humanitarian efforts. On January 27, it was discovered that the south pier was more damaged than it initially appeared and could not be used safely. It had previously been used by one ship at a time, gingerly unloading one container at a time.

NAVFAC engineers utilized the damage assessment to determine the remaining capacity of the damaged South Pier and determine the ability of the structure to support emergency cargo operations. Only very limited cargo operations were allowed with significant load restrictions established and with some areas of the pier closed completely. Therefore, task force efforts to restore port operations at PPAP focused on construction of cargo and equipment staging areas along the shoreline until repairs could be made to the uncollapsed length of the South Pier.

Construction of Staging Areas and Temporary Port Facilities

The task force quickly initiated construction of four landing beaches of rubble and sand fills along the inner shoreline of PPAP. Three landing beaches, Red Beach #1, #2, and #3, were completed by the task force, and a fourth landing beach, APN Rubble, was constructed by APN (Fig. 7-43). These beaches were built to receive humanitarian, military, and commercial cargo. A landing beach, referred to as White Beach, was also constructed near the Varreaux Terminal, primarily for military purposes.

Later, the task force also coordinated the installation of a 122 m by 30 m (400 ft. by 100 ft.) barge at Red Beach 3 (which was renamed APN Blue #1) and a second barge (APN Blue #2) on February 27 (Fig. 7-43). These barges were held in place with spud piles and were connected to land with 300-ton capacity ramps. They were utilized as temporary wharves to accommodate unloading of the bulk of the commercial and emergency relief cargo arriving in Haiti. Two 230-ton crawler cranes were used to load and unload cargo.

With these barges and staging areas in place, transportation of cargo into Haiti was restored. However, most of the cargo ships calling on Haiti had a draft exceeding the relatively shallow water depths (about 6 m to 8 m [20 to 25 ft.]) alongside the barges. Thus, it was necessary to transfer the cargo offshore from the larger ships to vessels with a smaller draft. These vessels were then anchored alongside the barges and offloaded onto trucks driven onto the barges.

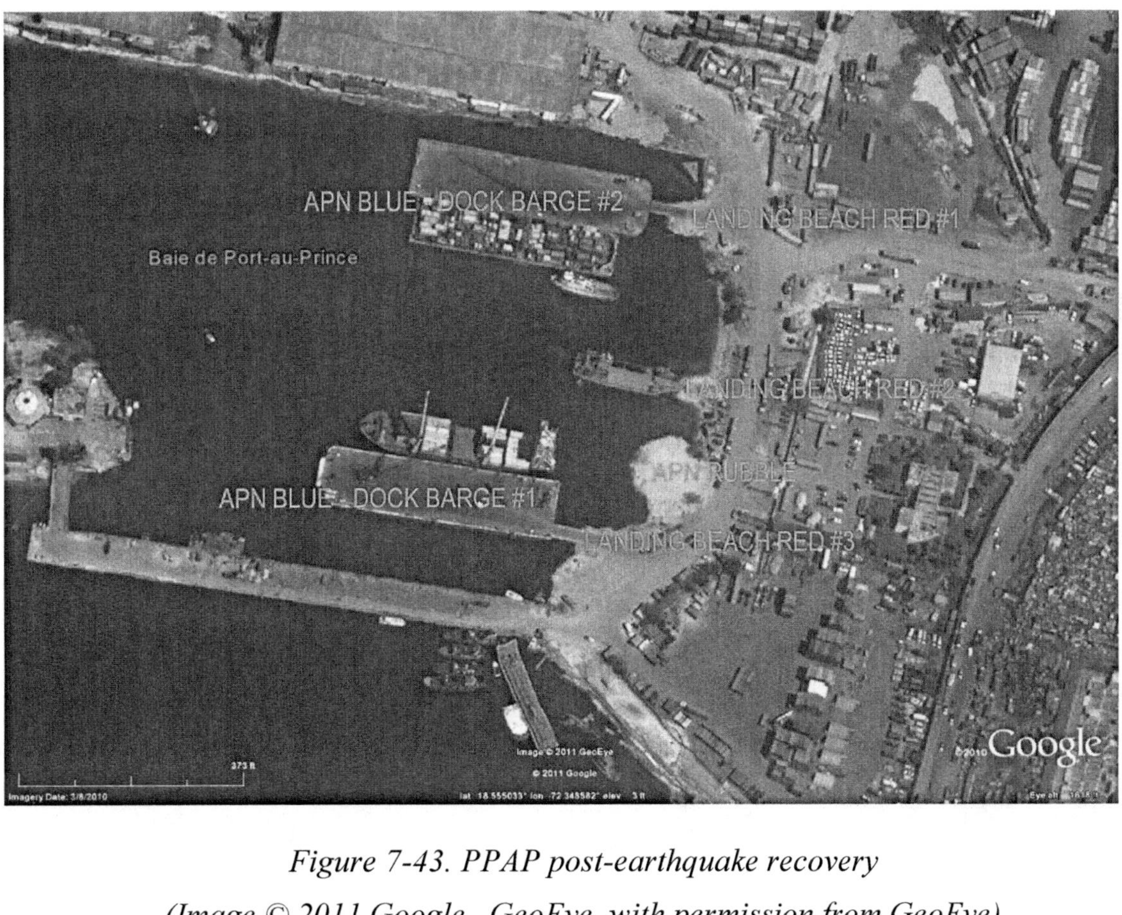

Figure 7-43. PPAP post-earthquake recovery
(Image © 2011 Google, GeoEye, with permission from GeoEye)

These shipping and cargo handling operations enabled commercial shipments, humanitarian relief, and emergency response/recovery cargo to reach the port and the people of Haiti. With the first barge in place at APN Blue #1, cargo throughput into the port reached 200 to 250 TEU per day and increased to up to 600 TEU per day toward the end of February. The subsequent deployment of the APN Blue #2 on February 27 further increased the port's cargo throughput.

South Pier Repairs

Repairs to the South Pier were completed by personnel from UCT-1, 544th Engineer Dive Team, and NMCB7. These repairs were undertaken to restore the pier's load carrying capacity to its pre-earthquake levels, but not to seismically upgrade the pier. The repairs consisted of two primary methods.

1. The first method was utilized where the shear core of the pile remained intact at the pile to superstructure connection. For this condition, an epoxy mortar repair was used to protect the remaining pile, reinforcing and re-establishing the cross section.
2. The second method was utilized in areas where the pile to deck connection had failed completely in shear. For this condition, a cast-in-place, reinforced concrete splice or cap was constructed to re-establish the connection between the remaining pile and the superstructure. Vertical reinforcing bars were embedded in the underside of the concrete deck with epoxy adhesive.

The task force divers formed a reinforcing bar cage around the damaged piles, embedded the vertical rebar into the deck, and placed transverse reinforcing around the vertical reinforcing bars. Wooden forms were used to cast the new pile caps. As many as seven piles per day were repaired. The repairs were completed by the end of March 2010—less than three months after the earthquake.

North Wharf Reconstruction

Reconstruction of the North Wharf will be a major effort. It will require removal of the North Wharf's structural elements on the bottom of the shipping channel and replacement of the fills along the waterfront as well as the wharf structure, cranes, and infrastructure that would service the wharf (buildings, power systems, access roads, etc.).

White Beach Staging Area

The U.S. Navy Seabees also constructed a staging area near the Varreaux Terminal, referred to as White Beach. The area was primarily used to receive cargo essential for U.S. military support of repair activities and humanitarian efforts. It was also used by emergency relief agencies to transport people, equipment, and supplies essential to relief activities.

The White Beach area included a large floating pier consisting of three 100-ton modules interconnected and deployed within two weeks of the U.S. Navy Seabees' arrival in Haiti. Cargo was transported as a Ro/Ro operation (Fig. 7-44). The area also included an underwater pipeline for bulk liquid transfer of up to 4.5 million liters (1.2 million gallons) per day and a camp for U.S. military personnel.

Thor Marine Oil Terminal

As of the end of March 2010, all damaged pipelines at the Thor Terminal had been repaired and all broken valves had been replaced. Temporary repairs of damage at Tank 5 are underway. There will reportedly be no repair of Tank 6. In addition, a bid for construction of a new jetty has been submitted and is being evaluated before a final decision is made. No information is yet available about plans for replacement of the terminal's main office building.

Figure 7-44. White Beach staging area
(Photo courtesy of Nason McCullough)

Varreaux Break Bulk and Marine Oil Terminal

During the reconnaissance it was observed that new fill was being placed where the terminal had settled during the earthquake. It was also observed that rubble from damaged buildings in Port-au-Prince was being used to reclaim additional land for future expansion of the terminal, which will reportedly include a new port, additional jetties, and a new terminal for accommodating container and break bulk cargo transport and storage.

CINA Cement Plant

The plant, pier, and Ro/Ro ramp were operational shortly after the earthquake. The remaining damaged segments of the conveyor were removed. New fill material was brought in a few days after the event to restore the ground level behind the Ro/Ro ramp due to settlement during the earthquake.

The Ro/Ro ramp was being used to transport humanitarian aid into Haiti.

Le Moulins d'Haiti Flour Mill

Given that the building completely collapsed during the earthquake, the flour mill was not in operation, and it was noted that it would take a year or more to rebuild the mill. The bulkheads were operational shortly after the earthquake, with fill placed in areas of settlement during the earthquake.

The east bulkhead was used to transport humanitarian aid into Haiti. The bulkhead was in service about 10 days after the earthquake and was used to transport about 105 TEU per day.

7.4 Major Observations and Recommendations

Significant damage occurred to port, marine oil terminals, and the airport during the earthquake. Damage was attributed to liquefaction of backfill and foundation soils and collapse of structures due to strong ground shaking. The waterfront and airport facilities were vital for post-earthquake humanitarian and recovery efforts.

The airport operation was restored within a day; however, all operations were controlled outside as the control tower and terminal were condemned following the earthquake. The damage to waterfront facilities was extensive and required time and significant effort to construct the temporary beaches at PPAP. The majority of the facilities needed to be replaced, which may take several years to accomplish. Some cargo was routed through smaller private ports, but overall the importation of humanitarian and recovery cargo was hampered by the lack of adequate and resilient waterfront structures.

7.5 Acknowledgments

Our investigative efforts were aided by numerous people, and we are grateful for their assistance. We wish to acknowledge Alex Holsteinson and Ann Hauge for providing numerous contacts within the government and private sector; Jean Evans Charles and Nacesse Pierre at APN; Gregory Mevs, Fritz Mevs, and Richard Lebrun at WinGroup and Terminal Varreaux; Donald Toussaint at Thor Marine Oil Terminal; Brian Crowder, Josh Tomblin, Mathew Hahne, Lowell Schrader, Michael Teates, Tracy Bower, Richart Kahler, Timothy Menzie, John Singler, Joe Stavish, and Mark Bradford of the U.S. military; and Werner Knecht and Clarence Fremont at Les Moulins d'Haiti.

8. Hospitals and Medical Services

The earthquake caused major structural damage to numerous health care facilities. Several hospitals collapsed, numerous hospitals were closed due to loss of all functionality, and many hospitals could only provide reduced services, often in tents outside the building. All facilities lost electrical power for various periods of time, which significantly disrupted hospital functions. Additionally, the commercial supply of water was interrupted in many facilities.

8.1 Description of System

The World Health Organization (WHO)/Pan American Health Organization (PAHO) has compiled a list of healthcare facilities in Haiti. In its response update of March 3, 2010, PAHO noted that before the earthquake there were "594 primary health care centers (including centers without beds and hospitals with beds), 30 reference communal hospitals (30 to 60 beds each), six centers for integrated diagnostics, 10 department hospitals (150 beds each); and three university hospitals (1500 beds in total). Before the earthquake, numerous non-governmental organizations (NGOs) had a presence in Haiti that provided health care services, training for health care providers, and advice to Haitian health care administrators. For example, PAHO had a core staff of 52. In support of the emergency, PAHO sent an additional 60 professionals with expertise in disaster management, logistics, epidemiology, communicable disease control, water, and sanitation, among others.

Figure 8-1 provides an overview of the status of Haiti's health care system before the earthquake including the average population covered per hospital for the various Departments of Haiti. The average population covered per primary health care facility before the event is shown for the various Departments of Haiti in Figure 8-2. Figure 8-3 shows Port-au-Prince health care facilities by category (hospital, primary health, other).

8.2 Overview of System Performance

In the response update of March 3, 2010, PAHO indicated that many hospitals "sustained damage but continue to function, often with the help of NGOs." PAHO's initial assessments indicated that some 49 health facilities had structural damage that rendered them unsafe.

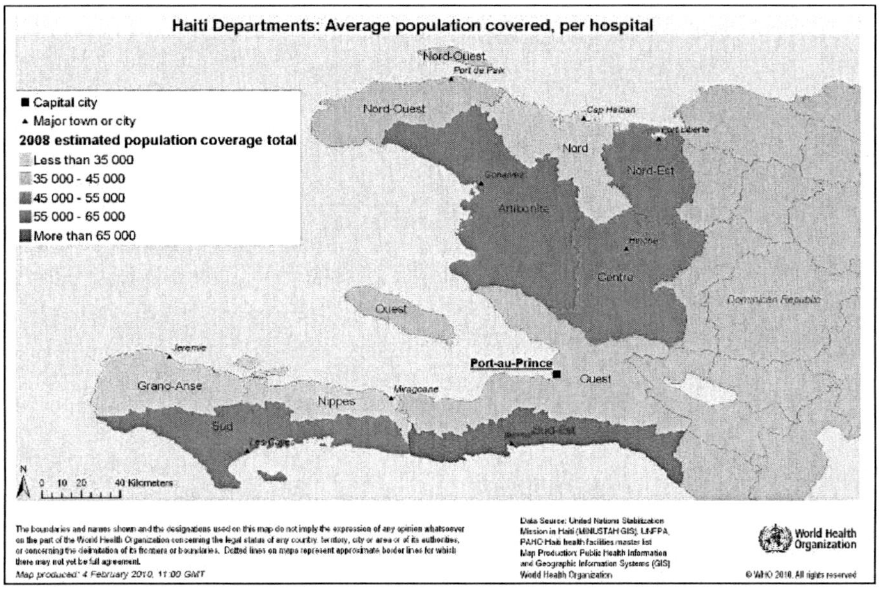

Figure 8-1. Average population covered per hospital by Haiti departments. (Reproduced by permission from the World Health Organization)

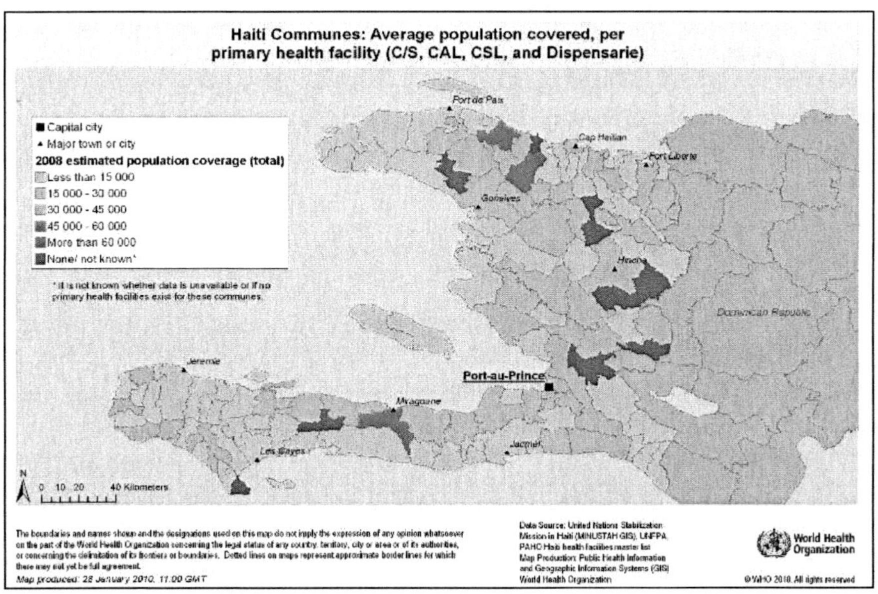

Figure 8-2. Average population covered per primary health facility by Haiti Communes (Reproduced by permission from the World Health Organization)

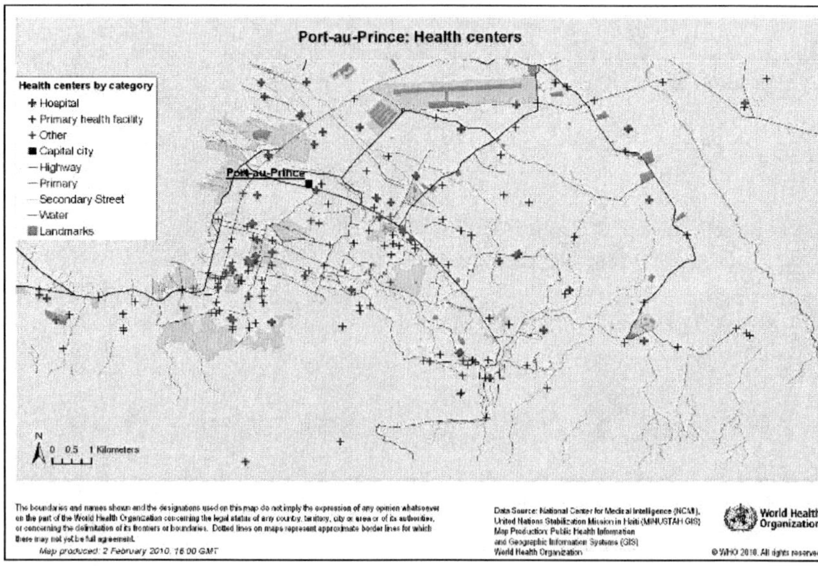

Figure 8-3. Port-au-Prince Health centers by category (Reproduced by permission from the World Health Organization)

In a special report dated February 16, 2010, PAHO stated that Haiti's Ministry of Health lost more than 200 staff members in the earthquake. Many of these fatalities occurred when the Ministry of Health building collapsed.

In the aftermath of the earthquake, there were 91 functioning hospitals in Haiti. The operational status of medical facilities around Port-au-Prince two weeks after the earthquake is shown in Figure 8-4.

Of the 91 functioning hospitals in the country, 59 were in the metropolitan Port-au-Prince area. In this metro area, the functioning facilities included four public hospitals, 34 NGO or private-run hospitals, and 21 field hospitals. Fifty-six of the 59 hospitals had surgical capacity. The operational status of medical facilities in Port-au-Prince two weeks after the earthquake is shown in Figure 8-5.

It was reported that patients were reluctant to enter hospital buildings, due to fear of collapse during aftershocks. Consequently, almost all healthcare services were provided in tents, even 7 weeks after the event. Most facilities (public and private) were providing services to patients free of charge. At least one private hospital was not able to pay its staff due to a lack of income. Both public and private hospitals were eager for written recommendations on seismic retrofit and upgrade. They intended to use the recommendations, along with data on patients served for free, to solicit additional funding from NGOs.

The details of the healthcare facilities investigated for this report are presented in the next section.

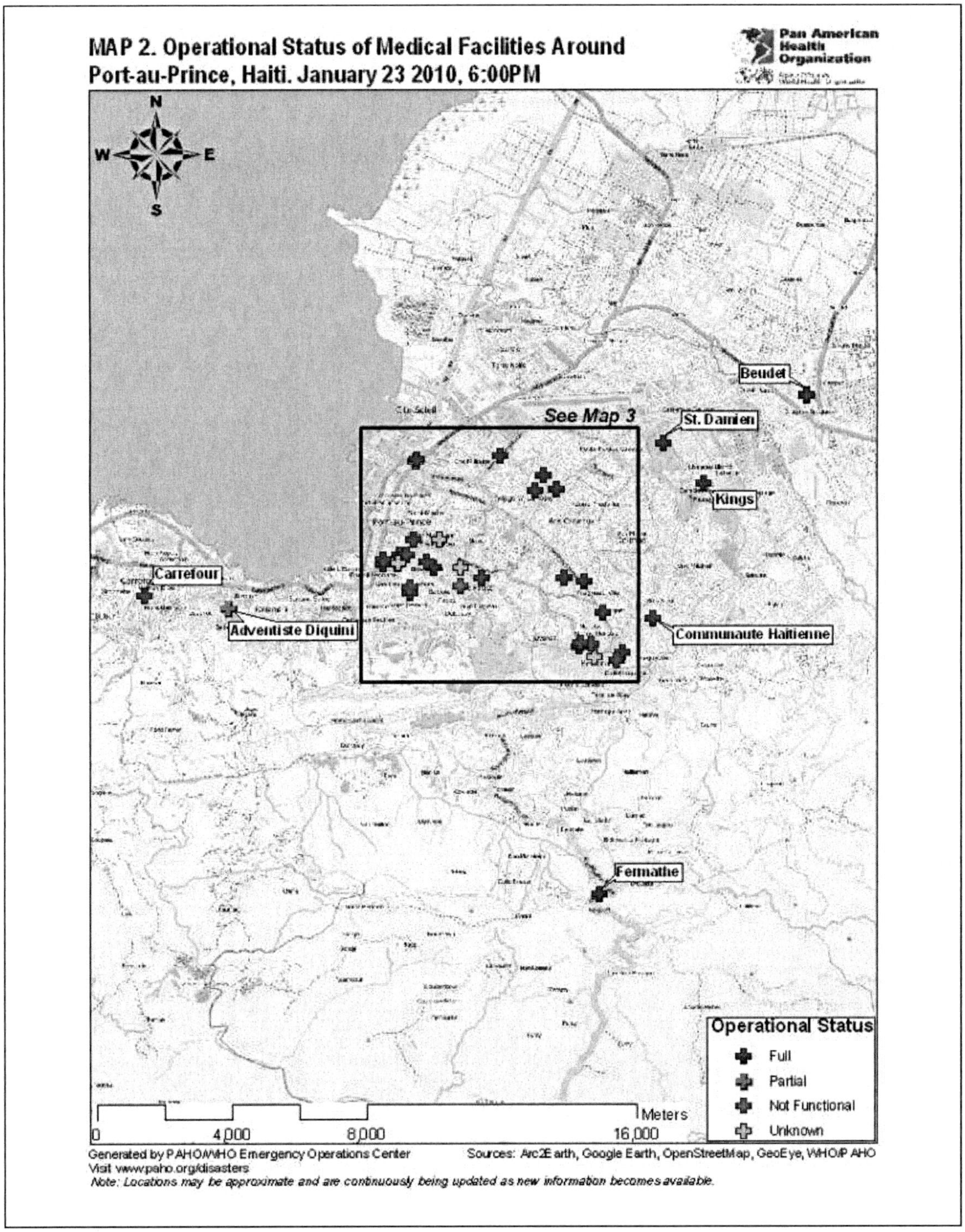

Figure 8-4. Operational status of medical facilities around Port-au-Prince (Reproduced by permission from the World Health Organization)

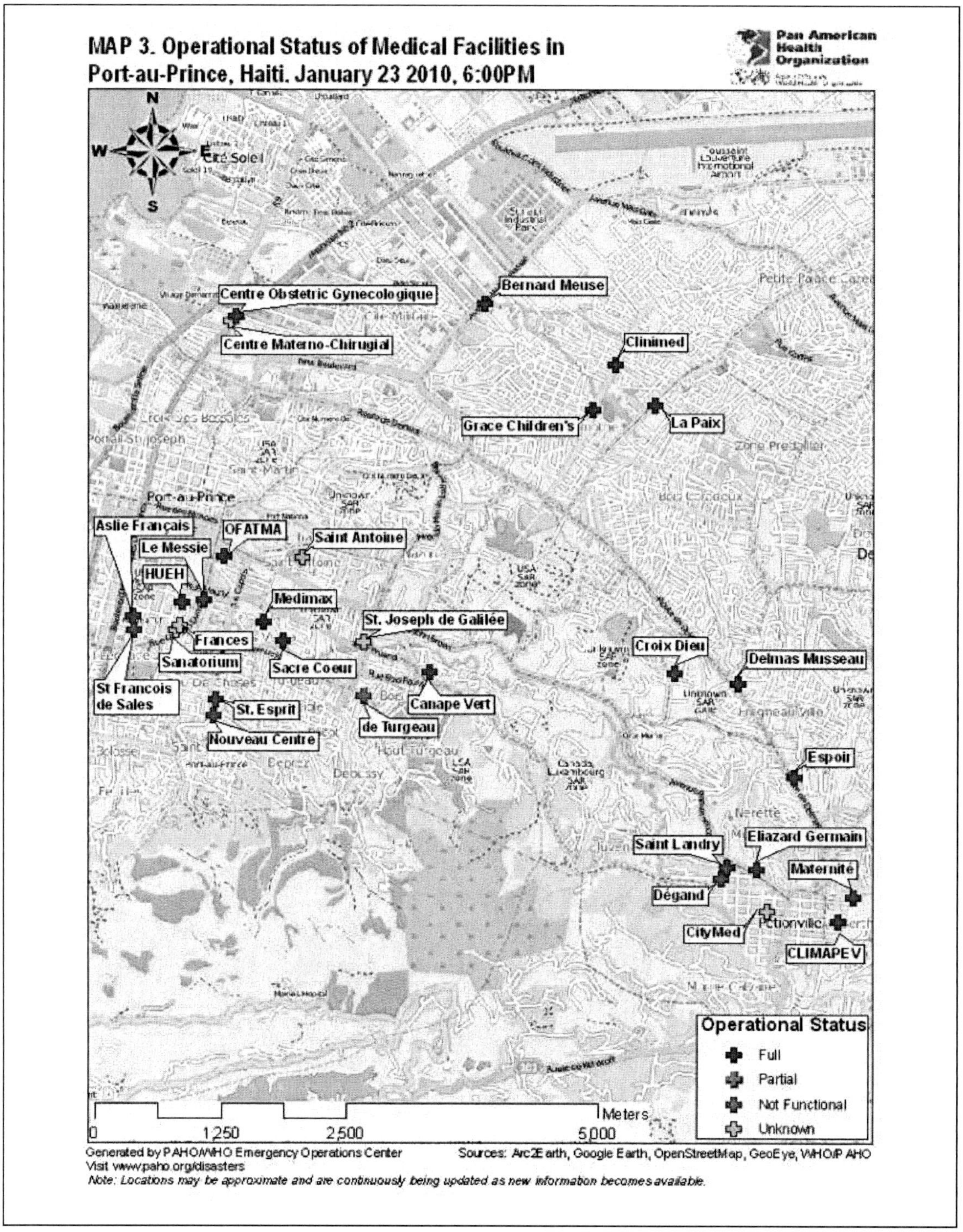

Figure 8-5. Operational status of medical facilities in Port-au-Prince (Reproduced by permission from the World Health Organization)

Grace Children's Hospital

The Grace Children's Hospital was located in Delmas, in the eastern part of metropolitan Port-au-Prince (18.5597°N, 72.3019°W) and was funded by various NGOs.

Prior to the earthquake, about 250 patients (in- and outpatients) were treated daily. Due to the earthquake, there was one patient fatality onsite and five staff fatalities offsite. Seven weeks after the event, about 125 outpatients were treated daily.

Grace Children's Hospital consisted of several structures located on two campuses and separated by a street. The older campus contained one- and two-story structures that were built in the 1950s with reinforced concrete and unreinforced masonry infill walls. In 1973 the older campus became a hospital to provide maternity and pediatric care and to treat children with tuberculosis (TB) and HIV. It had 60 beds.

On the older campus, severe damage occurred in the two-story structure containing the main administration building and the TB/HIV clinic and the two-story structure containing the maternity and pediatric wards. In each structure there were many failures at beam-column joints due to non-ductile lateral reinforcement detailing (Fig. 8-6 and 8-7), numerous collapsed walls (Figure 8-8), and numerous locations of severe pounding damage (Figure 8-9). These structures were unoccupied and will be demolished.

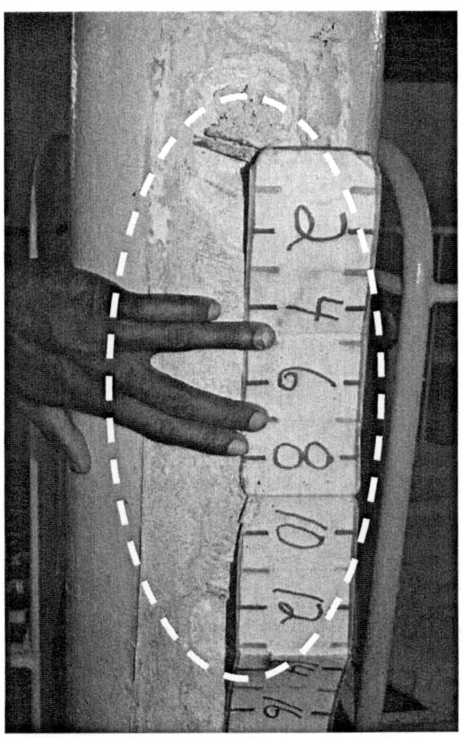

Figure 8-6. Column lateral reinforcement spacing of 30 cm (12 in.), Grace Children's Hospital (Photo by Mark Pickett)

Figure 8-7. Beam-Column failure due to non-ductile detailing, Grace Children's Hospital (Photo by Mark Pickett)

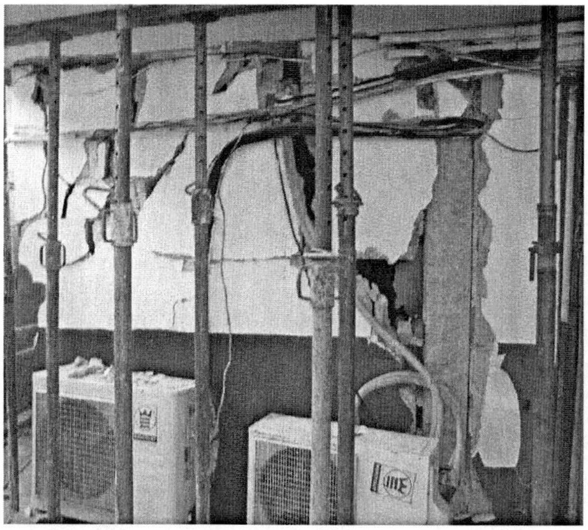

Figure 8-8. Shear cracks in unreinforced masonry infill wall and unanchored chiller units, Grace Children's Hospital (Photo by Mark Pickett)

Figure 8-9. Pounding damage, Grace Children's Hospital (Photo by Mark Pickett)

The newer campus of Grace Children's Hospital contained two structures rigidly connected by a lobby structure. The structure containing the ophthalmology clinic was built in 2009; the structure containing the reproductive health clinic was built in 1999. Due to pounding, there were several minor cracks in the walls at the rigid connections with the lobby structure.

Les Centres GHESKIO

Les Centres GHESKIO was a large organization, founded in 1982, to provide clinical treatment, research, and training in the specialties of tuberculosis, HIV/AIDS, and other infectious diseases. GHESKIO had facilities at two locations. One facility, identified as INLLR, was located just south of downtown Port-au-Prince (18.5378°N, 72.3485°W). The other facility, identified as IMIS, was located on the northern side of the city (18.5706°N, 72.2509°W), near the airport and U.S. Embassy.

GHESKIO receives about 100,000 patient visits annually. Because it receives funding from various NGOs, GHESKIO provides all services and medications free of charge.

The INLLR facility was located on a 12,000 m^2 (3-acre) campus, consisting of eight one- and two-story buildings built between 1950 and 1980, with a total area of about 4650 m^2 (50,000 ft^2). The buildings were constructed of reinforced concrete with unreinforced masonry infill walls. At this site, the water table was so high that all the commodes on the ground level utilized pumps to flush the wastewater. Several of the buildings had settlement problems before the earthquake. As a result of the event, a clinic building settled about 15 cm (6 in.) over a span of 3m (10 ft.) and the administration building settled about 30 cm (1 ft.).

The IMIS facility was located on a 32,000-m^2 (8-acre) campus, consisting of a clinic building and a separate research/medical center building. The soil at the site had been compacted. The one-story research center building consisted of the Warren D. Johnson Jr. Medical Center (built 2004) and the Laboratoire Rodolphe Merieux (built 2009). The laboratory structure was about 44 m by 18 m (144 ft. by 59 ft.) and was connected to the medical center structure by a hallway of the same height (12.8 m by 6 m, 42 ft. by 20 ft.) and a lobby structure of the same height (31 m by 6 m, 102 ft. by 20 ft.). The medical center consisted of two structures, each shorter than the lobby, yet each rigidly connected to the lobby. One structure was about 5 m by 9.4 m (16.4 ft. by 31 ft.); the other was about 24.6 m by 29 m (81 ft. by 95 ft.).

There was significant damage to several columns and two roof beams in the lobby area. This was caused by pounding due to the relative motion between the tall columns (386 cm, 12.7 ft.) in the lobby and the shorter columns (256 cm, 8.4 ft.) in the medical center. The columns were separated by a gap of insufficient size (2.5 cm, 1 in.) (see Fig. 8-10).

Also due to pounding, there was a full-wall-height crack at the joint between the laboratory hallway and the lobby. The shear wall at the south end of the lobby had large shear and flexural cracks from the foundation to the parapet level (Fig. 8-11).

Figure 8-10. Damage caused by pounding of different height columns, IMIS, GHESKIO (Photo by Mark Pickett)

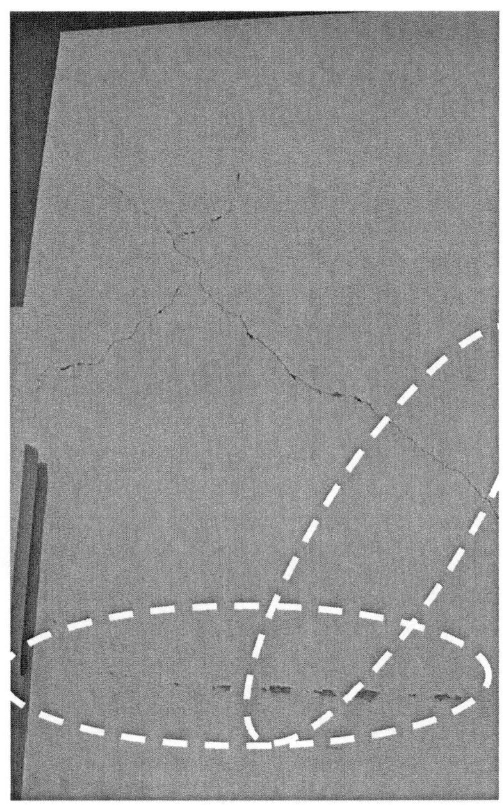

Figure 8-11. Shear and flexural cracks in shear wall, IMIS, GHESKIO (Photo by Mark Pickett)

Figure 8-12. Electrical conduit, false ceiling grid and HVAC ducts, IMIS, GHESKIO. (Photo by Mark Pickett)

In the IMIS laboratory structure, there were cracks at wall corners and wall-column joints and shear cracks in one external masonry wall. One external wall was slightly out of plumb. Several ceiling tiles and hanging-light fixtures suffered damage (Fig. 8-12).

Hôpital St. Joseph

The Hôpital St. Joseph was located in Haut Port-au-Prince (Prince (18.5350°N, 72.3281°W) and funded by NGOs. The facility had one emergency room and about 40 beds.

Immediately after the earthquake, all staff members and patients (about 30 to 40 patients) were evacuated. The earthquake caused two onsite fatalities. Reconstruction started 2 weeks after the event. Seven weeks after the earthquake, the facility was in the process of reconstruction and was not providing health care to patients onsite.

The rear portion of the facility consisted of a three-story structure (built 1992-1999), constructed of reinforced concrete with unreinforced masonry infill walls. The front portion of the facility was a three-story structure (built 2000), constructed of reinforced concrete with unreinforced masonry infill walls. At the time of the earthquake, an additional two stories were being constructed on the front portion, using reinforced concrete. The reinforcement grid of the new back wall fell onto the roof of the older rear structure, causing the two fatalities.

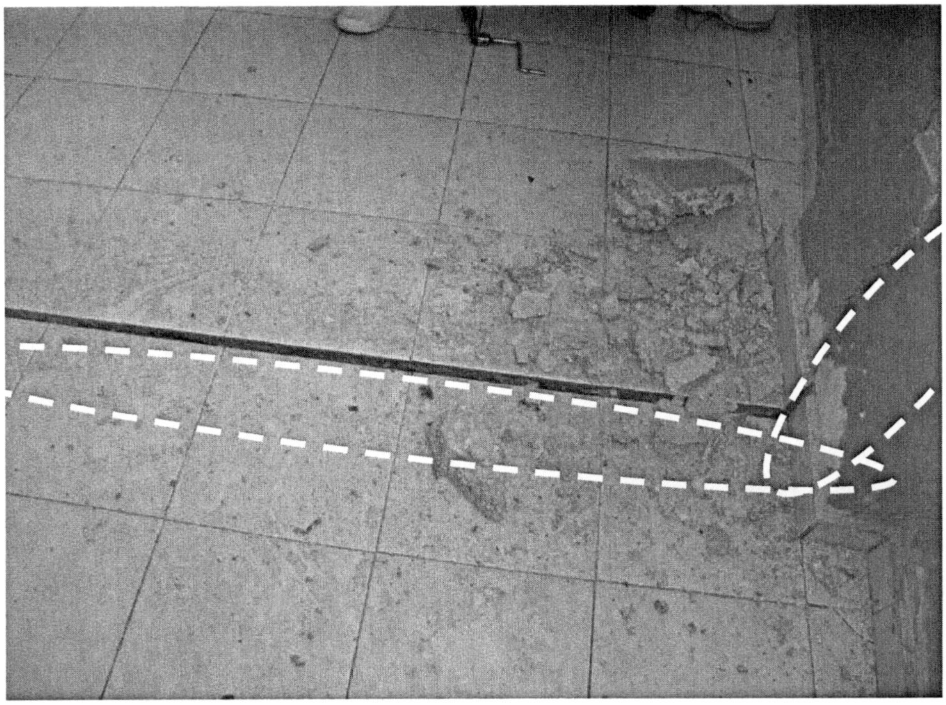

Figure 8-13. Pounding damage, Hôpital St. Joseph (Photo by Mark Pickett)

On the first level, pounding damage was observed between beams, columns, and floors of the rear and front structures (Fig. 8-13). Significant cracks were observed in the unreinforced masonry infill walls, particularly in the rear structure.

In the damaged portion of St. Joseph's newer front structure, damaged partition walls were being replaced with unreinforced concrete block walls, concrete columns reinforced with #4 (13 mm, 0.5 in.) smooth bars, and #3 (9.5 mm, 0.375 in.) lateral ties spaced at 13 cm (5 in.).

Sanatorium de Port-au-Prince #1

The Sanatorium de Port-au-Prince #1 was a large public facility located at 18.539°N, 72.335°W. Before the earthquake this facility served about 200 to 300 in-patients daily with a staff of about 230. Seven staff members were injured in the earthquake. All patients and staff were evacuated immediately following the earthquake. Seven weeks after the event, about 30 in-patients and 50 outpatients were being treated in tents, daily.

The sanatorium facility consisted of an L-shaped three-story building, with a wooden roof (initially constructed in about 1942, with an addition constructed later). The older portion of the structure was constructed with rubble-rock-and-brick columns, unreinforced concrete block beams, and unreinforced concrete block infill walls (Fig. 8-14). The mortar in the masonry walls was soft and brittle and could be easily removed with a metal hand-punch. In the older portion, many columns had cracks that extended deep into the rubble-rock interior and over at least half the column height. The newer portion was constructed of reinforced concrete columns and beams with unreinforced masonry infill walls. In the newer portion, many infill walls had large full-thickness shear cracks, and many walls had collapsed. This structure was not functioning 7 weeks after the event.

At the rear of the L-shaped building, there was a newer one-story administrative office structure. A second level was being added at the time of the earthquake. This structure was constructed of reinforced concrete beams and columns with unreinforced masonry infill walls. The reinforcement in numerous beams consisted of twisted wire strands (Fig, 8-15). The column lateral ties at several failed beam-column joints were #2 (6 mm, 0.25in.) spaced at 20 cm. This structure was completely destroyed.

Figure 8-14. Unreinforced masonry beam and infill wall, Sanatorium Hospital (Photo by Mark Pickett)

Figure 8-15. Twisted wire strand in beam, Sanatorium Hospital (Photo by Mark Pickett)

Centre Hospitalier du Sacré-Coeur (CDTI)

The Centre Hospitalier du Sacré-Coeur was a private facility located in Haut Port-au-Prince at 18.5373°N, 72.3313°W. It was recently renamed CDTI. The facility had four emergency rooms, two surgery rooms, one endoscopy room, and 18 beds. Before the earthquake, about 1000 patients were served annually, including an average of 1.5 surgeries per day. Before the event, the staff included 50 medical professionals and 25 other staff members. The facility was evacuated immediately after the earthquake. After the earthquake, almost all patient care was performed in tents outside the structure (Fig. 8-16). This included a daily load of about 60 to 75 in-patients, 10 surgeries, and 10 dressing applications. The emergency room inside the damaged structure was used to treat about 60 to 75 outpatients per day.

Figure 8-16. Patient care in tents, CDTI Hospital (Photo by Mark Pickett)

CDTI consisted of a three-story structure (built around 2006), constructed of reinforced concrete shear walls (front transverse wall thickness, 35.6 cm [14 in.]) with unreinforced masonry partition walls. The rear of the structure formed a semi-circle in the transverse direction. The structure was about 10 m by 42 m (33 ft. by 138 ft.).

On the first level, there were numerous cracks in the unreinforced masonry infill walls. In the semi-circular rear of the structure, partial-height unreinforced masonry walls restrained the lower half of some columns, causing short-column-effect damage (Fig. 8-17). In the semi-circular rear, near the top beam-column joints of some damaged columns, the #3 lateral tie spacing was about 16 cm (6.3 in.).

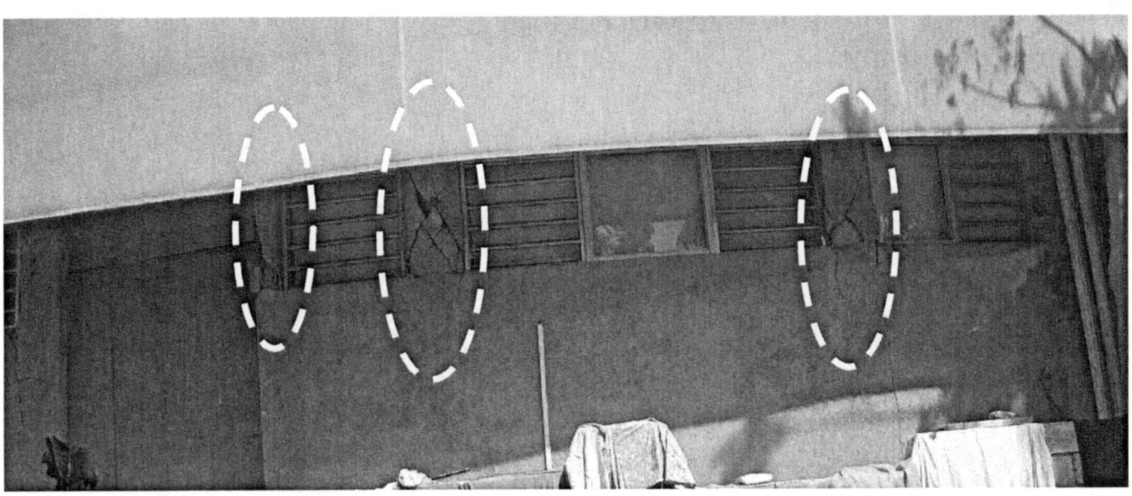

Figure 8-17. Short-column-effect damage at CDTI (Photo by Mark Pickett)

Hôpital de l'Universati de Haiti (HUEH)

The Hôpital de l'Universati de Haiti (HUEH) was a large public facility located in Port-au-Prince at 18.5402°N, 72.3395°W. The facility had about 700 beds and separate buildings for emergency rooms, surgery rooms, outpatient clinics, internal medicine, and maternity-obstetrics-gynecology (Fig. 8-18).

Before the earthquake, about 1000 outpatients and 100 emergency room patients were served daily. All patients and staff were evacuated immediately after the earthquake. There were no injuries from the earthquake. After the event, almost all patient care was performed in tents outside. This included a daily load of about 1000 to 1500 patients during the first month; however, some patients had to be turned away due to a lack of staff and safe facilities. The facility's staffing levels 7 weeks after the event; physicians, 20%; nurses, 80%; administration, 50%, other staff, 66%. The daily load at this time was 250 to 300 in-patients and 500 to 900 outpatients.

Figure 8-18. Map of buildings at Hôpital de l'Universati de Haiti (HUEH) (Photo by Mark Pickett)

Figure 8-19. Overturned partial height partition walls, Hôpital de l'Universati de Haiti (HUEH) (Photo by Mark Pickett)

The HUEH facility consisted of numerous two- and three-story structures (built from 1934 onward) constructed of reinforced concrete with unreinforced masonry partition walls (Fig. 8-18). At least three buildings had been evacuated and would be demolished (surgery, pediatrics, and internal medicine). Numerous safety threats were posed by the overturning of partial height unreinforced masonry partition walls (Fig. 8-19).

Hôpital du Canape-Vert

The Hôpital du Canape-Vert was a 60-bed private facility (including two intensive care unit beds, seven maternity beds, two emergency room beds, and three surgery beds) in Haut Port-au-Prince (18.5382°N, 72.3119°W).

Before the earthquake, this facility served on a daily basis about 40 in-patients, 200 outpatients, and seven emergency room patients. At the time of the earthquake, there were about 40 patients and 25 staff members in the facility. There were no onsite injuries due to the earthquake.

Immediately after the earthquake, the entire facility was evacuated. It re-opened for surgery two days later. All services were functioning 33 days after the event.

The Canape-Vert facility consisted of several two-story detached buildings, built of reinforced concrete with unreinforced masonry infill walls. One structure (built 1960) contained administrative offices and about 22 beds, some of which were occupied within a few days of the event. This structure had a few cracks in the masonry walls. An 11-bed structure (built 2004-06) was unoccupied 7 weeks after the event, but wall crack plastering was underway so that it could be reoccupied. A laboratory building (built

1990) was rigidly connected to a structure (built 2004-06) containing about 13 beds. The laboratory structure remained occupied because it had only minor wall cracks. However, there was significant pounding damage at the beam-column joints connecting the laboratory building to the newer structure and also to a stairwell. The newer structure was unoccupied due to damage at several other beam-column joints caused by non-ductile reinforcement detailing. The surgery building (built 2004-06) was also unoccupied due to similar damage at beam-column joints.

Clinque-Hôpital Le Messie

The Clinque-Hôpital Le Messie was a 14-bed private facility in Bas Port-au-Prince (18.5389°N, 72.3389°W). It also contained one emergency room, two dental care rooms, and one eye care room.

Before the earthquake this facility served about 14 in-patients and 50 outpatients daily. Immediately after the earthquake, the entire facility was evacuated, and patient care was provided in an open courtyard outside of the building. The facility was about 25 percent functional within 1 week and fully functional within 2 weeks of the event. Seven weeks after the event, 50 outpatients were treated inside daily.

The facility consisted of a main two-story building (built 1995-96) of reinforced concrete with unreinforced masonry infill walls.

Hôpital Saint Esprit

The Hôpital Saint Esprit was a 35-bed private facility in Bas Port-au-Prince (18.5388°N, 72.3389°W). It also contained one emergency room, one major surgery room, and one minor surgery room. Before the earthquake this facility served about 35 in-patients and 100 outpatients each day. Immediately after the earthquake, the entire facility was evacuated. The earthquake caused injuries to 12 patients/staff members. On the first day after the event, 900 patients were treated outside of the building in tents. Within the first week, the full pre-event range of patient care was provided in tents. Seven weeks after the event, St. Esprit was fully functional, but very few functions were being performed inside the building, mainly due to patient fear of aftershocks.

The facility consisted of a two-story structure (built about1960) and an addition containing the surgery rooms (built 1998); both were built of reinforced concrete with unreinforced masonry infill walls. The only earthquake-related damage observed was a few minor cracks in several infill masonry walls.

MediMax Hospital

The MediMax Hospital was a private facility in Haut Port-au-Prince (18.53039°N, 72.3369°W), located one block from Hôpital Saint Esprit.

Before the earthquake, MediMax served about 75 in-patients and 100 outpatients each week. It also had an operating room and a delivery room. There were no event-related injuries. Immediately after the earthquake, 15 in-patients were evacuated. Seven weeks after the event, the facility was fully functional. However even though there was very little earthquake damage, only seven outpatients were being treated each day because patients were afraid of aftershocks.

The MediMax facility consisted of a single four-story building (built 2000), constructed of reinforced concrete with reinforced masonry infill walls. The only earthquake-related damage observed was a few cracks in several infill masonry walls. In one wall, a shear crack went completely through the thickness of the wall.

At the time of the earthquake, a seven-story reinforced concrete hospital building was under construction in front of the current MediMax structure. It was designed to contain 30 patient rooms, three operating rooms, and one emergency room. This building collapsed completely, but no injuries or casualties occurred. It was observed that several lower-level columns were reinforced with #5 axial bars (0.625 in., 16 mm) and #3 lateral ties (0.375 in., 9.5 mm). It was noted that several columns with tie spacing near the beam-column joint of about 10 cm (4 in.) performed in a ductile manner, but that several columns with tie spacing near the joint of about 21 cm (8.5 in.) failed in a brittle manner.

Hôpital de Turgeau

The Hôpital de Turgeau was a three-story facility located in Port-au-Prince Prince (18.5322°N, 72.3237°W), just across the street from the high-rise Digicel building. The weak first story of the hospital, which was made of reinforced concrete, collapsed. It was observed that some column axial bars were 13 mm (0.5 in.) with lateral ties 6 mm (0.25 in.) spaced at 15 cm (6 in.). The structure had been evacuated, and there was no patient care its vicinity.

Turgeau Hospital Health Center SODEC

The Turgeau Hospital Health Center SODEC was a multi-story facility located in Port-au-Prince at 18.5339°N, 72.3245°W. The entire reinforced concrete structure collapsed. It was observed that some column axial bars were 13 mm (0.5 in.) and 16 mm (0.625 in.) with no lateral ties. The floor beam stirrups were 6 mm (0.25 in.) spaced at about 30 cm (12 in.).

Centre de Santé des Srs. de Ste. Anne St. Antoine

The Centre de Santé des Srs. de Ste. Anne St. Antoine was a small two-story private facility located in Port-au-Prince at 18.5259°N, 72.3351°W. It had an open first level of reinforced concrete columns. Many of these columns failed in a brittle manner at the top beam column joint, due to non-ductile detailing. It was observed that some column axial smooth bars were 9.5 mm (0.375 in.) with lateral ties 6 mm (0.25 in.) spaced at 15 cm (6 in.). The second level had unreinforced masonry infill walls. The structure had been evacuated, and there was no patient care near the structure. Pharmaceuticals were scattered around on the first level floor. A four-story primary school building was adjacent to the medical facility. The school building had numerous collapsed unreinforced masonry infill walls. It had been evacuated.

OFATMA Annex

The OFATMA Annex was a two-story military facility located in the central square area near the Presidential Palace in Port-au-Prince (18.5435°N, 72.3350°W). It was a reinforced concrete structure with unreinforced masonry infill walls. The walls suffered much damage.

CityMed/Petion Ville Hospital

The CityMed/Petion Ville Hospital was an eight-bed facility located in Petion Ville in the southeast part of metropolitan Port-au-Prince (18.5140°N, 72.2869°W) and was funded by NGOs.

Before the earthquake about 50 outpatients were served per day. There were no onsite injuries due to the earthquake. Seven weeks after the event, about 100 to 150 outpatients were served daily for free.

The facility consisted of a single two- to three-story structure, constructed of reinforced concrete with unreinforced masonry infill walls. There were only a few cracks in the walls and the facility was nearly 100 percent functional.

Centre Hospitalier Eliazar Germain

The Centre Hospitalier Eliazar Germain was a private facility located in Petion Ville in the southeast part of metropolitan Port-au-Prince (18.5172°N, 72.2887°W).

The facility had five patient rooms with 10 beds per room. Seven weeks after the event, about 100 outpatients were served each day for free.

The Germain facility consisted of a single three-story structure (built 2006-07), constructed of reinforced concrete with unreinforced masonry infill walls. On the roof, it was observed that the reinforcement extending from the columns consisted of #5 axial bars (16 mm, 0.625 in.) with #2 lateral ties (6 mm, 0.25 in.) at a spacing of 15 to 20 cm (6 to 8 in.). There were only a few cracks in the walls, and the facility was nearly 100 percent functional.

St. Landry Hospital

The St. Landry Hospital was located in Petion Ville, in the southeast part of metropolitan Port-au-Prince (18.5160°N, 72.2898°W), and funded by NGOs.

The facility had four in-patient rooms, with several emergency and surgery rooms. Before the earthquake, the normal daily load was three to four in-patients and 40 to 50 outpatients. There were no onsite injuries due to the earthquake. Because this facility was fully functioning less than 12 hours after the earthquake, 43 surgeries were performed in the first few days working around the clock. Seven weeks after the event, outpatients were being treated for free.

The St. Landry facility consisted of a single four-story structure (completed in November 2009) constructed of reinforced concrete. According to the owner, the structure was designed and constructed according to American seismic building codes. Only a few minor surface cracks on a few exterior and interior walls were observed. The facility was virtually 100 percent functional.

Hospital Espoir

The Hospital Espoir facility was located in Delmas, in the east part of metropolitan Port-au-Prince (18.5419°N, 72.2783°W) and funded by NGOs. The facility had a capacity for 50 in-patients.

Before the earthquake, the normal daily load was 20 in-patients and 30 outpatients. There were no onsite injuries from the earthquake. Immediately after the earthquake, the facility provided daily care to about 60 in-patients and 250 to 300 outpatients. Seven weeks after the event, some patient care was provided inside the complex and some was provided in tents outside.

The Espoir facility was a three-story L-shaped complex (front part built in 1993, rear part built in 2007), constructed of reinforced concrete and unreinforced masonry infill walls. Several cracks were observed where the old and new buildings pounded together and at joints between walls and ceilings.

Hospital La Croix Dieu, Complex Medico-Chirugical

The Hospital La Croix Dieu, Complex Medico-Chirugical, was located in Delmas, in the east part of metropolitan Port-au-Prince (18.5385°N, 72.2956°W). La Croix Dieu was funded by various NGOs and by factories for their low-income employees.

La Croix Dieu was the headquarters of a complex of private facilities that included five hospitals and six outpatient clinics (DASH), located throughout the Port-au-Prince metropolitan area (Fig. 8-20). One of these other facilities, St. Charbel (a 24-bed, three-story facility located in Delmas at 18.5467°N, 72.3047°W collapsed completely causing 40 to 50 fatalities.

La Croix Dieu also functioned as an outpatient clinic. Before the earthquake its normal daily load was 100 outpatients. There were no onsite injuries due to the earthquake. During the first month after the earthquake, the facility provided free outpatient care to about 400 patients per day. Seven weeks later, about 100 outpatients per day were served, paid for by various NGOs and by factories for treatment of their low-income employees. At the event, the staffing level was at about 50 percent.

The facility consisted of three-story structure (built 2000), constructed of reinforced concrete and unreinforced masonry infill walls. A few small cracks were observed at wall-column or wall-ceiling interfaces.

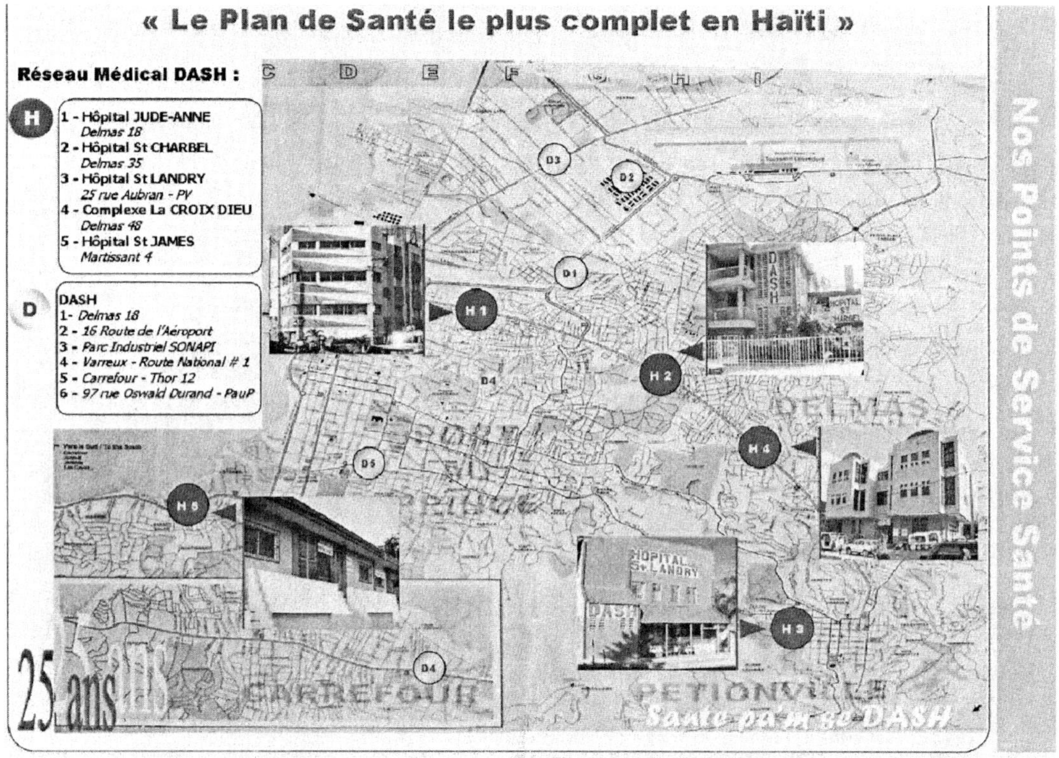

Figure 8-20. Map of the complex of hospitals and outpatient clinics, La Croix Dieu Hospital (Photo by Mark Pickett)

Hôpital Delmas Musseau

The Hôpital Delmas Musseau was a private facility located in Delmas, in the east part of metropolitan Port-au-Prince (18.5318°N, 72.2908°W).

Before the earthquake, the normal daily load was six in-patients, 20 outpatients, and two to three emergency room patients. There were no onsite injuries due to the earthquake. For the first five days following the earthquake, the facility provided free care in tents outside the structure to about 500 patients. Seven weeks after the event, about 10 to 15 outpatients per day were being served, but patients were still afraid to enter the building even though only a few cracks were observed.

The Delmas Musseau facility consisted of a two-story structure (built 1990), constructed of reinforced concrete and unreinforced masonry infill walls. A few small cracks were observed at wall-column or wall-ceiling interfaces. Seven weeks after the event, an emergency exit ramp was being constructed from the second level to the courtyard.

Hôpital Universitaire de la Paix

The Hôpital Universitaire de la Paix was a large public facility located in Delmas, in the east part of metropolitan Port-au-Prince (18.5569°N, 72.2983°W).

The facility had about 100 beds and a pre-earthquake daily load of 200 to 300 outpatients. Seven weeks after the event, care was provided inside the complex to about 500 patients per day.

The hospital facility consisted of one- and two-story structures (built 2002-03), constructed of reinforced concrete and unreinforced masonry infill walls. The structures were rigidly connected together by covered walkways around an open-air courtyard. Pounding caused numerous cracks at the solid joints between the walkways and the structures.

8.3 Electrical Power Systems: Commercial, Emergency, and Solar

The facilities investigated reported that the loss of commercial electrical power from Electricite' d'Haiti (EDH) occurred immediately after the earthquake. These facilities reported that their emergency generators were operated for many hours, before the earthquake, due to the intermittent nature of the EDH power supply.

Inverter and Batteries Systems

Most of the facilities investigated relied on a back-up system consisting of an inverter using between eight to 36 6V batteries (see Fig. 8-21). None of the batteries observed were adequately restrained against lateral motion.

Figure 8-21. Inverter and un-restrained batteries, Clinque—Hôpital Le Messie (Photos by Mark Pickett)

Figure 8-22. Manually operated breakers for the emergency power system, MediMax Hospital (Photo by Mark Pickett)

Automatic Transfer Switches

None of the facilities investigated had any automatic transfer switches (ATS). They manually switched from normal to emergency power (Fig. 8-22).

Emergency Power Generators – Stationary Generators, Fuel Pumps, and Fuel Tanks

All of the facilities investigated had emergency generators. Several facilities operated their generators 24 hours a day, or at least between 8 a.m. and 4 p.m., as part of their normal operations to ensure an uninterrupted power supply for medical procedures and patient care. A few generators had rubber vibration isolation pads (Fig. 8-23). One generator had spring isolators (Fig. 8-24). Most generators were unanchored. Two roof-top generators slid almost 2 m (6.5 ft.). Only a few generators were on concrete pads. Some generators were refueled by hand with 19 L (5 gal.) portable containers. The facilities that had automatic fueling systems reported no problems with the fuel pumps, day tanks, or inground fuel tanks. One rooftop day tank was unanchored (Fig. 8-25). Two hospitals had above ground saddle-supported fuel tanks. No leakage was observed in these tanks or piping systems. Figure 8-26 shows one fuel tank supported on an A-frame. Table 8-1 presents data on the emergency power systems of the facilities investigated.

Figure 8-23. Emergency generator with rubber vibration isolator pads, St. Landry Hospital (Photo by Mark Pickett)

Figure 8-24. Spring vibration isolators of emergency generator, Hôpital Universitaire de la Paix (Photo by Mark Pickett)

Figure 8-25. Unanchored day tank for emergency generator, St. Landry Hospital (Photo by Mark Pickett)

Figure 8-26. Saddle-supported fuel tank mounted on A-frame, OFATMA Annex (Photo by Mark Pickett)

Table 8-1. Summary of Hospital Emergency Power Systems

Facility	When Regain EDH Power	Normal Daily Operation of Emergency System	Emergency Generators	Emergency Generators Anchored	Gen. Isolated	Gen. on Concrete Pad	Refueling System	Inverter with Number of Batteries
Grace Children's	Never relied upon	8AM–4 PM	1-300KW	No	No	No	Above ground unanchored	Yes, 4PM—8AM
GHESKIO INLLR IMIS	IMIS Never relied upon	INLLR As needed IMIS=24 hours	INLLR=750KW IMIS=200KW IMIS=400KW	? ? ?	? ? ?	? ? ?	? ? ?	? ? ?
St. Joseph	?	Many hours	2-300KV	Slid 2m	No	No	In ground	?
CDTI	Never relied upon	24 hours	2-257KV 1-115KV	No No	No No	Yes Yes	In ground 19000 L	?
HUEH	6 weeks	Many hours	4	No	No	Only 1	In ground	?
Canape-Vert	2 weeks	7PM—7AM	2-175KW 1-275KW	?	?	?	?	?
Le Messie	3 weeks	As needed	1-14KVA	No	No	No	By hand	Yes, with 16 batteries
St. Esprit	15 days	>12 hours	1-17KW	No	No	No	By hand	Yes, with 12 batteries
MediMax	2 weeks	As needed	1	No	No	No	By hand	Yes, with 24 batteries
CityMed	6 weeks	As needed	1-40KW	No	No	No	By hand	?
Eliazar Germain	Never relied upon	24 hours	1-300KV	No	Rubber Pads	Yes	In ground	Yes
St. Landry	Never relied upon	24 hours	1-220KV	No	Rubber pads	No	In ground	Yes, with 12 batteries
Espoir	Never relied upon	24 hours	1	1 bolt Bent over	No	Yes	750 L 200 gal. Tank supporting generator	Yes, with 36 batteries
La Croix Dieu	?	Daytime as necessary Nighttime Inv+batteries	1 Plus portable on trailer	No	No	No	By hand	Yes
Musseau	Never relied upon	24 hours	1-12KW	No	No	No	Above ground unanchored	Yes
La Paix	?	As necessary	2-500KV	No	Springs	No	Above ground unanchored	Yes, with 24 batteries

Emergency Power Generators – Portable Generators

None of the facilities reported the use of portable hand-carried generators.

Solar Power Generators – Solar Water Heaters

St. Landry Hospital had an unanchored solar water heater on the roof (Fig. 8-27). No leakage was observed. Hospital Espoir had four sets of rooftop unanchored photovoltaic panels (Fig. 8-28). No problems were observed with either set of panels. La Paix hospital had photovoltaic panels on top of light poles (Fig. 8-29).

Figure 8-27. Solar water heater, St. Landry Hospital (Photo by Mark Pickett)

Figure 8-28. Solar photo-voltaic electrical panels, Hospital Espoir (Photo by Mark Pickett)

Figure 8-29. Solar panel and rigidly connected structures, Hôpital Universitaire de la Paix (Photo by Mark Pickett)

8.4 Water Systems

The PAHO Special Report of February16, 2010, reported that safe drinking water was being distributed at the rate of 5 liters per day per person to more than 780,000 people in 300 sites across Port-au-Prince, Leogane, and Jacmel. The same PAHO report mentioned that about 50 water tanker-trucks were providing water to health care facilities. These trucks were collecting water from three major pumping stations where the water was chlorinated as it was pumped from 50 m underground. PAHO Situation Report #20 reports that a permanent water aqueduct had been installed in Buen Samaritano and that the water uptake capacity was increased in General Melenciano Hospital. Both hospitals were along the Dominican Republic border.

Many health care facilities reported that normally the external piped-in water supply was used for patient bathing and commode flushing but that potable water was obtained from trucks. Some facilities received about 3000 L (800 gal.) per week via truck, in addition to the external piped-in water. Other facilities received significantly more water via truck, in addition to water from the external sources. St. Landry received 3000 L every 2 days. Delmas Musseau received 19,000 L (5000 gal.) every 2 weeks. Grace Children's Hospital received 6000 L each week. The large public hospital, HUEH, received about 5000 L every 2 days. CityMed hospital used only water supplied by truck.

Many (but not all) facilities filtered and/or purified water for human consumption, regardless of source—external, trucks, or wells (Fig. 8-30). At the Hospital Saint Esprit, chlorine was added to the potable water daily. A team of Spanish military personnel provided a filtering system that was used to provide water for drinking and patient bathing at HUEH. The system consisted of activated carbon, sand filtration, and chlorination (Fig. 8-31).

Many facilities relied solely on their own onsite wells. These facilities included GHESKIO-IMIS, CDTI, St. Joseph, Espoir, and La Paix.

GHESKIO-INLLR utilized the external water supply and its own onsite wells.

The La Croix Dieu facility collected rainwater from the roof and stored it in a cistern. This water was sufficient for the daily needs of this outpatient clinic.

The Hôpital du Canape-Vert was the only facility to report that internal water pipes were broken due to wall movement and cracking. This facility had two concrete reservoirs. One was in good condition, and the other had a crack about 1 m from the top.

An emergency water bladder had been setup in the courtyard of the OFATMA Annex (Fig. 8-32).

In the Sanatorium de Port-au-Prince #1 facility, it was apparent that the older L-shaped structure had functioned without electrical power and without a water-in-pipes system for a long time before the earthquake. The newer one-story structure did have water pipes in its floors and ceilings.

Many facilities had unanchored water tanks on the roof, but some had rooftop tanks that were well anchored (Fig. 8-33).

Figure 8-30. Water purification system, Hôpital St. Joseph (Photo by Mark Pickett)

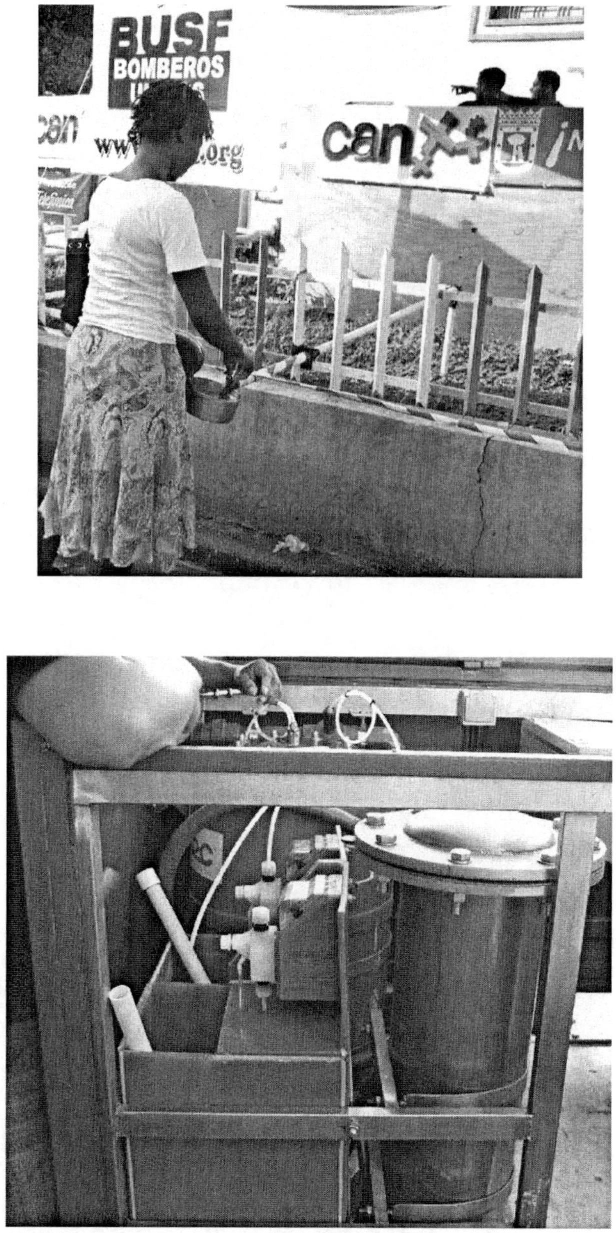

Figure 8-31. Patient drawing water from water purification system, HUEH (Photo by Mark Pickett)

Figure 8-32. Water bladder, OFATMA Annex (Photo by Mark Pickett)

Figure 8-33. Anchored water tanks, unanchored chiller units and columns' reinforcement extensions, Eliazar Germain Hospital (Photo by Mark Pickett)

The La Paix hospital had a reinforced-concrete water tank on its roof (Fig. 8-34).No cracks were observed in the tank. Cracks were, however, observed at several locations in the support structures of the reinforced concrete water tanks at HUEH hospital. These tanks were empty and unused before the earthquake (Fig. 8-35).

8.5 Wastewater Systems

No facility reported any problems with the either the external or internal wastewater systems. However, at the GHESKIO INLLR campus, the water table was so high that all the ground-level commodes used pumps to flush the wastewater.

8.6 Communication Systems: External and Internal

In Situation Report #20, PAHO reported that about 43,000 radios were distributed to people in Port-au-Prince in an effort to communicate key health messages (e.g., water sanitation, handling patients, etc.) via Haiti public service broadcasts.

All facilities investigated reported that TELECO landline and all cellular phone communications were lost immediately following the earthquake. Many facilities had not regained landline communications 7 weeks after the event. However, most facilities reported that cellular communications were restored within about 2 weeks.

Figure 8-34. Water tank and unanchored roof-top air chiller units, Hôpital Universitaire de la Paix (Photo by Mark Pickett)

Figure 8-35. Cracks in cross beam of tower supporting water tank, HUEH (Photo by Mark Pickett)

In most hospitals, staff members communicated via cell phones. CDTI utilized two-way phones (walkie-talkies) and internal intercom systems. However, many facilities had neither.

The Espoir facility had a radio communications antenna mounted on the roof for external communications (see Fig. 8-36). The Eliazar Germain hospital had a roof mounted satellite dish for external communications. There was no apparent damage to this dish (Fig. 8-37).

8.7 Heating, Ventilating, and Air Conditioning Systems

In most of the facilities investigated, there were individual air conditioning (AC) units in various rooms, including Le Messie, Esprit, MediMax, Canape-Vert, La Croix Dieu, Germain, St. Landry, Espoir, La Paix, and Grace Children's Hospital. Most of the chillers for these AC units were mounted on the roof (Fig. 8-38, 8-33, and 8-34). Some were mounted on external walls or on the ground (Fig. 8-8 and 8-39). None of the chillers observed were properly anchored. Some wall mounted AC in-room units at La Croix Dieu had fallen to the floor.

Air ducts were observed only in GHESKIO-IMIS, Canape-Vert, and CDTI facilities. Some ducts were ruptured by falling debris in Canape-Vert. No problems were reported with the ducts in CDTI and GHESKIO-IMIS.

There were individual room fans in the St. Joseph and Delmas Musseau hospitals.

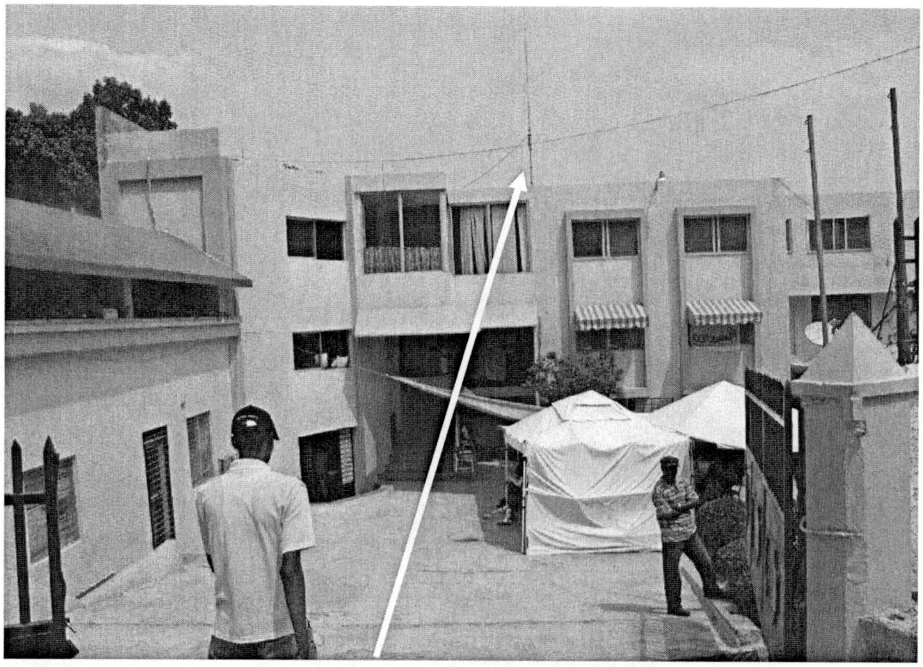

Figure 8-36. Radio communications antenna and patient treatment tents, Hôpital Espoir (Photo by Mark Pickett)

Figure 8-37. Roof mounted communications satellite dish, Eliazar Germain Hospital (Photo by Mark Pickett)

Figure 8-38. An unanchored, roof-mounted air conditioning chiller unit at Eliazar Germain Hospital (Photo by Mark Pickett)

Figure 8-39. Typical unanchored wall-mounted air conditioning chiller unit at CDTI Hospital (Photo by Mark Pickett)

8.8 Natural Gas and Propane Gas Systems

Bottled propane gas cylinders were used for cooking in all the facilities investigated. The Le Messie facility used propane and electrical power for cooking. As there is no natural gas system in the region, no facility reported using natural gas for heating, cooking, or sterilization.

8.9 Patient Transportation Systems

Only two hospitals had elevators—CDTI (not in use) and St. Landry (in use). In the other facilities, patients were transported from floor to floor via stairs or ramps. No motor vehicle ambulances were observed at any of the facilities.

8.10 Medical Gas Systems

The CDTI and St. Landry hospitals did have medical gas systems with piping systems that extended through the internal walls. There were no ruptures of these medical gas lines. In the other facilities investigated, the medical gases were supplied to patients via cylinders on portable carts. None of the cylinders observed had adequate restraint (Fig. 8-40). Several facilities reported problems with overturning gas cylinders.

Figure 8-40. Unrestrained medical gas bottles, Hôpital Saint Esprit (Photo by Mark Pickett)

8.11 Critical Equipment

In the facilities reliant on EDH commercial power, the inverter-with-batteries systems would power some critical equipment, upon the loss of EDH power. One hospital reported that six manometers fell and broke. Another reported that all but one of its manometers fell and broke.

8.12 Fire Suppression Systems

None of the facilities investigated had fire sprinkler systems. Hôpital Saint Esprit had no fire suppression equipment. All the other facilities had numerous hand-held fire extinguishers (Fig. 8-41). No fire hose standpipes were observed.

Figure 8-41. Hand-held fire extinguisher, Hôpital Espoir (Photo by Mark Pickett)

8.13 Hazardous Material

Numerous facilities reported problems from biohazard material falling and spilling on the floor. Many facilities had numerous corpses on their grounds the first few days after the earthquake. At least two facilities reported that there were no biohazard problems.

8.14 Security Systems

Most facilities reported that they had no problems with looters. However, Canape-Vert reported that looters took pharmaceuticals and items left behind by evacuated patients.

8.15 Emergency Preparedness, Response, and Recovery

The WHO/PAHO Emergency Operations Center (EOC) in Washington, D.C., went into full response mode upon notification of the earthquake. The WHO/PAHO EOC mobilized members of the Regional Health Emergency Response Team to Port-au-Prince and set up a primary logistics base about a 1.5 hour drive east of Port-au-Prince in Jimaní, Dominican Republic (DR). The Haitian Ministry of Health, with the support of PAHO/WHO, developed an intermediary plan to guide the rebuilding of the health care system. PAHO/WHO is supporting the Haitian government in developing a national plan for disability and rehabilitation.

Immediately following the event, the response focused on rescuing people buried in the rubble and providing immediate emergency care for trauma patients. After about 1 month, the focus of the response changed to post-operative care, follow-up of surgery patients, basic primary health care, and the prevention of the spread of infectious diseases.

PAHO estimates that the January12, 2010 earthquake resulted in more than 217,000 fatalities, more than 300,000 people suffering injuries, more than 500,000 people leaving the Port-au-Prince area for outlying departments, more than 97,000 houses being destroyed, and more than 188,000 homes being damaged. See Table 8-2.

Table 8-2 Summary of Affected Population by Department in Haiti

Department	Dead	Missing	Injured	Affected Families	People in Shelters	Displaced People	Houses Destroyed	Houses Damaged
Artibonite	9		19	783	45	162,509		
Centre	12		39	4279		90,997		
Grande-Anse						55,167	80	703
Nippes						33,350	257	4022
Nord						13,531		
Nord-Est						8500		
Nord-Oest						45,862		88,762
Oest	211,607	369	299,803	91,523	1,215,790	32,253	23,029	
Sud						25,532		
Sud-Est	441	14	711	39,334	21,242		9,621	30,478
Total	**212,069**	**383**	**300,572**	**135,919**	**1,237,077**	**467,701**	**32,987**	**123,965**
Revised Totals								
Feb. 19, 2010							97,294	188,383
March 3, 2010	217,000+					500,000+		

(Source: PAHO Special Reports; February 16, February, 19, and Update of March 3, 2010)

In a February 16, 2010 Special Report, PAHO reported that there were now 10 organized settlements with a total of 42,569 people and 315 spontaneous temporary settlements that were occupied by 468,075 people (91,112 families)". It also reported that "sanitation was becoming a major concern at many of the temporary sites." During the first month, international NGOs distributed latrines to the settlements in an attempt to achieve a ratio of one latrine for every 50 people, with an ultimate goal of one latrine per 20 people within 6 months.

In the March 19, 2010 "Situation Report #30," the United Nations Office for the Coordination of Humanitarian Affairs (OCHA) reported that some 302,000 children had been displaced to other departments, with an additional 720,000 children affected by the earthquake but remaining in their home communities. Of this number, 309,500 children were currently living in spontaneous temporary settlement sites that lack basic social services. Additionally, OCHA stated that there were 460 sites (including sub-sites for the largest settlements) with a total population of 1,170,000 in the Port-au-Prince area. PAHO reported that more than 270 international organizations provided health care services in at least 15 communities.

In response to needs expressed by the Haitian Government, WHO/PAHO procured at least 10 mobile clinics. During the first month, 87 mobile units and outpatient centers were opened for the treatment of severely acute malnutrition. An additional 52 centers were scheduled to be opened to treat patients with malnutrition during the second month after the event.

At least eight patient care sites were set up along the Dominican Republic border by international agencies. During the first month, these sites focused on surgical activities.

On the third day after the earthquake, a PAHO team set up a site in Jimaní, DR, which served more than 3000 patients in the first few days. After the first month, these sites focused on rehabilitative care, with more than 1200 patients receiving care each day. At one of these border sites, the Buen Samaritano Hospital, a permanent water aqueduct was installed, while at the General Melenciano Hospital in Jimaní, DR, the water uptake capacity was increased.

None of the facilities investigated reported having an emergency response plan.

8.16 Conclusions, Observations, and Recommendations

Health care facilities in earthquake prone regions need to be retrofitted to provide ductile beam-column joints and/or shear walls so that the health care functions can be performed in the aftermath of an earthquake.

Hospitals need to have automatic transfer switches so that emergency power can be immediately provided to critical equipment and systems when an earthquake causes the loss of normal electric power. Health care facilities must ensure that their emergency generator(s) are adequately anchored, vibration-isolated, and provided with slack/flexible connections (electrical, fuel) so that they can remain operational to supply sufficient electrical power for the normal functioning of the hospital, thus facilitating the adequate treatment of injured patients.

Landline and cellular systems often fail in the first few days of a disaster. Therefore, health care facilities must have redundant communication systems. Interior and exterior communications need to be provided by two-way radios (walkie-talkies) and radio telephones. These devices must have rechargeable batteries and chargers that are connected to the emergency power system.

8.17 Acknowledgments

The author is indebted to the following people who took time to provide valuable information, thus making it possible to reduce losses in future earthquakes:

Dr. Justine Crowley, CDTI Hospital

Dr. Gilbert Jean Charles, Hôpital Universitaire de la Paix

Dr. Elrick Metayer, Hôpital Le Messie

Dr. Jean Danthan Morose, MediMax Hospital

Dr. Jean William Pape, GHESKIO

Pere Calixte Hilaire, Hôpital St. Joseph

Dr. Nathalie Pierre Louis, Hôpital St. Landry

Dr. Jacques Geneus, Hôpital Delmas Musseau

Dr. Rudolph Magloire, Grace Children's Hospital

Dr. Ken Hover, P.E., Cornell University

8.18 References

OCHA, United Nations Office for the Coordination of Humanitarian Affairs, Haiti Earthquake, Situation Report #30, 19 March 2010.

PAHO, Pan American Health Organization, Special Report: Update on the Health Response to the Earthquake in Haiti, 16 Feb. 2010.

PAHO, Pan American Health Organization, Emergency Operations Center Situation Report #20, 16 Feb. 2010.

PAHO, Pan American Health Organization, Emergency Operations Center Situation Report #21, 19 Feb. 2010.

PAHO, Pan American Health Organization, "Earthquake in Haiti, The Response of the Pan American Health Organization/World Health Organization," Update March 3, 2010.

Some of the figures and information for this chapter were retrieved on February 23, 2010 from the following Web pages with permission from the World Health Organization:

- www.who.int/hac/crises/hti/maps/haiti_population_per_hospital_district_4feb2010.jpg
- www.who.int/hac/crises/hti/maps/haiti_population_per_primary_health_facility_d_4feb2010.jpg
- www.who.int/hac/crises/hti/maps/Haiti_Earthquake_Portauprince_hospitals_20100202.jpg
- www.who.int/hac/crises/hti/maps/AroundPaPHospitalsOperationalStatus.jpg
- www.who.int/hac/crises/hti/maps/PaPHospitalsOperationalStatus.jpg

Index

Page numbers followed by f and t indicate figures and tables, respectively.